SpringerBriefs in Applied Sciences and Technology

PoliMI SpringerBriefs

AF820943

Springer, in cooperation with Politecnico di Milano, publishes the PoliMI SpringerBriefs, concise summaries of cutting-edge research and practical applications across a wide spectrum of fields. Featuring compact volumes of 50 to 125 (150 as a maximum) pages, the series covers a range of contents from professional to academic in the following research areas carried out at Politecnico:

- Aerospace Engineering
- Bioengineering
- Electrical Engineering
- Energy and Nuclear Science and Technology
- Environmental and Infrastructure Engineering
- Industrial Chemistry and Chemical Engineering
- Information Technology
- Management, Economics and Industrial Engineering
- Materials Engineering
- Mathematical Models and Methods in Engineering
- Mechanical Engineering
- Structural Seismic and Geotechnical Engineering
- Built Environment and Construction Engineering
- Physics
- Design and Technologies
- Urban Planning, Design, and Policy

http://www.polimi.it

Ornella Iuorio

From Mass Prefab to Mass Customization

Modern Methods of Constructions from Experimentation to Manufacturing

POLITECNICO
MILANO 1863

Springer

Ornella Iuorio
Department of Architecture, Built
Environment and Construction Engineering
Politecnico di Milano
Milan, Italy

ISSN 2191-530X ISSN 2191-5318 (electronic)
SpringerBriefs in Applied Sciences and Technology
ISSN 2282-2577 ISSN 2282-2585 (electronic)
PoliMI SpringerBriefs
ISBN 978-3-031-73326-0 ISBN 978-3-031-73327-7 (eBook)
https://doi.org/10.1007/978-3-031-73327-7

This Springer imprint is published by the registered company Springer Nature Switzerland AG
The registered company address is: Gewerbestrasse 11, 6330 Cham, Switzerland

Preface

The construction industry stands at a pivotal crossroads, where the demands of a rapidly urbanizing world intersect with the pressing challenges of climate change. As we grapple with a global housing crisis and the urgent need to reduce our environmental footprint, modern methods of construction (MMC) have recently gained prominence as a critical approach to address pressing global issues, including the housing shortage, environmental sustainability, and resilience.

This book embarks on a journey through the historical evolution of MMC since its first inception, up to the latest framework that articulates it into seven categories, going from the volumetric systems, through non-systematized industrialized components, up to those developed through the most disruptive technologies, which include additive manufacturing and robotic. The book delves into reviewing relevant developments in additive manufacturing that are quickly finding applications in the construction industry for their capacity to smartly develop high-performance systems. The book also explores how the integration of cutting-edge automated construction techniques, grounded in robotic manufacturing and assembly, from the earliest stages of building system design, has the potential to unlock new formal and technical paradigms, that are deeply connected to traditional scientific principles, and that remain still largely unexplored.

Large space is given to discuss the latest innovations embedded in prefab systems to provide low-carbon systems, advance circularity in the built environment, and retrofit the existing building stock. The text reflects upon the role played by research in the development of systems for mass manufacturing, but also for democratizing construction. It does it by discussing applied research and development, as well as by showing a selection of case studies that exemplify different levels of industrialization, applications, and role played by research. The book delineates, as one of the future lines for development, the enabling of deconstruction and reuse, discussing the opportunity to reuse lightweight façade systems to explore the carbon benefits of closing the loop.

The reflections included in this book are the result of twenty years of research in the field of prefabrication, which the author carried out in multiple countries. The author acknowledges all the researchers and colleagues who have collaborated over the years through a number of research projects. These projects, funded by public bodies, are: FARO, *Innovation and sustainability for building renewal. Best practice for Building Retrofit and Maintenance*; *Modular Housing, as response to the UK Housing Crisis*, Funded by ESRC LSSI IAA; *Digital Design and Fabrication*, Funded by Laidlaw; *Knowledge Transfer Partnership between University of Leeds and Ilke Homes*, funded by Innovate UK; *Prefabs sprouting: Modern Methods of Construction and the English Housing Crisis,* funded by ESRC*; UKRI Interdisciplinary Circular Economy Centre for Mineral-Based Construction Materials* (EP/V011820/1), funded by EPSRC; *Robotic construction of shell structures funded by EPSRC (2442110), and Exploration of Parametric Design and Robotic Assembly of Shell Structures*, Funded by EPSRC (2442114).

Milan, Italy | Ornella Iuorio

About This Book

This book provides an overview of the latest innovations in prefabrication. It analyses how digital, material, and process innovations are transforming the mass prefabrication of homes, school, and services into mass customization. It provides an understanding of available manufacturing processes, including distributed ownership of manufacturing, platform approaches, and robotics. It discusses how the integration of cutting-edge advanced construction techniques, coupled with robotic manufacturing and assembly, from the earliest stages of building system design, has the potential to unlock new formal and technical paradigms.

Investigating the impact of prefab in the context of climate emergency, it analyses the capacity and shortfall in delivering net zero emissions. It discusses the opportunities that modern methods of construction provide to enable the transition towards circular constructions, from reuse to retrofitting. Going beyond what has ever been said, demonstrates the importance of developing methodologies for capturing users' occupancy evaluation, as a means for understanding real performances, benchmarking indicators, and tuning systems to target the long-term well-being of the occupants.

Referring to a plethora of emblematic cases recently built and/or under construction, it demonstrates the importance of investing in research and development for optimizing construction systems, to reduce material use, facilitate lean construction, advance structural and environmental performances, and move towards circular systems to close the loop.

This book is aimed at practitioners, architects, technologists, researchers, and students in architectural and building engineering.

Contents

1 Historical Evolution of Prefabrication 1
1.1 The Origin of Prefabrication and Its Evolution 1
1.2 The Case of the Heavy Mass Production in Between the World Wars 5
1.3 Learning from Past Failures 8
1.4 Why Modern Methods of Construction Now? 10
1.5 The Newly Defined MMC Categories 12
1.6 Conclusion 13
References 14

2 Additive Manufacturing and Automation in Construction 17
2.1 Additive Manufacturing 17
2.1.1 3D Printing for Material Optimization 18
2.1.2 3D Printing for Customized Building Components 20
2.1.3 3D Printing for Building Production 23
2.2 Robotics in Construction 24
2.2.1 Robotic Platforms 25
2.2.2 End Effectors 28
2.2.3 Process Monitoring 29
2.3 Augmented and Virtual Reality for Design Exploration and Construction Monitoring 30
2.4 Conclusion 32
References 32

3 Engineering the Systems: The Role of Research and Development 35
3.1 The Case for Research and Development 35
3.2 Enabling Mass Manufacturing of Volumetric Lightweight Steel Houses Through Product Optimization 36

3.3 Development of Off-Site Constructions for Social Innovation 41
3.4 Development of a Platform Approach to Be Used in Multiple Contexts 45
3.5 Conclusion 50
References 50

4 The Role of Prefab in the Context of Climate Emergency 51
4.1 The Climate Emergency: Problem Definition 51
4.2 Development of MMC Systems to Achieve the Net Zero Targets 53
4.2.1 Net Zero Solutions for New Construction and Retrofitting the Built Environment 53
4.2.2 Methodologies for Assessing Performance and Sustainability of Low-Energy Buildings 55
4.2.3 Performance Evaluation of a Low-Energy Prefab Development in Oceanic Climate 56
4.3 Low-Embodied Carbon Potentials in a Life Cycle Perspective 59
4.3.1 Why Embodied Carbon Matters 60
4.3.2 Evaluation of Embodied Carbon in Wooden Based and Steel Based Prefab Systems 60
4.3.3 Future Opportunities to Reduce Embodied Carbon 63
4.3.4 Circular Systems: Can MMC Enable Circularity? 64
4.3.5 Circularity Streams Within the Prefab Industry 65
4.3.6 Circularity of Cold-Formed Steel Façade Wall Systems 66
4.3.7 Evaluating the Lifecycle Impact of Circular Lightweight Prefab Façade Systems 70
4.4 Conclusion 76
References 77

5 Emblematic Prefab Systems 81
5.1 WikiHouse 81
5.1.1 History 81
5.1.2 Research and Development 82
5.1.3 The System 82
5.1.4 Emblematic Examples 83
5.2 LeapHomes 85
5.2.1 History 85
5.2.2 Research and Development 85
5.2.3 The System 86
5.2.4 Emblematic Examples 86
5.3 Lightweight Steel Systems by Etex 87
5.3.1 History 87
5.3.2 Research and Development 88
5.3.3 The System 90
5.3.4 Emblematic Examples 91

5.4 Irondom 92
5.4.1 History 92
5.4.2 Research and Development 92
5.4.3 The System 93
5.4.4 Emblematic Examples 94
5.5 Automated Architecture 95
5.5.1 History 95
5.5.2 Research and Development 96
5.5.3 The System 97
5.6 Reframe Systems 98
5.6.1 The History 98
5.6.2 Research and Development 98
5.6.3 The System 99
References 100

Chapter 1
Historical Evolution of Prefabrication

Abstract The term Modern Methods of Construction (MMC) is typically used to denote forms of prefabrication. Although not a new phenomenon in and of itself, it has recently come to be recognized as an essential instrument in tackling the global housing shortage, and the ecological challenges, while providing resilience to disruptive phenomena. This chapter scrutinizes some of the main historical transformations of the concept of prefabrication, from initiations, through the major developments in the period around the World Wars, up to the most recent re-categorization within the framework of Modern Methods of Construction, which includes also the cutting-edge technologies which are disrupting the construction industry. By examining the historical context in which these methods developed, this chapter aims to bring the reader to reflect upon the potential they hold for transforming the built environment in response to contemporary challenges, while also providing a memory of the fall they have gone through in history. Finally, this section provides a synopsis of the MMC industry's recent rise to prominence as a self-described disruptor of conventional on-site building.

Keywords Portable prefabricated systems · Industrial production · Housing · Digital

1.1 The Origin of Prefabrication and Its Evolution

Portable prefabricated building systems have started to be produced systematically since the nineteenth century when the British and other European countries started to export them as a response to houses demand in America. The advent of Henry Ford's T model demonstrated that mass manufacturing could lead to high-quality objects, supporting in this way the development of prefabrication, so that at the start of the twentieth century many companies were producing high-quality and affordable prefabricated houses. Among them, one of the most important examples is the kit commercialized by Sears, Roebuck & Co. who developed a catalogue of high-quality precut material, that was shipped to the destination for being assembled with a series

O. Iuorio, *From Mass Prefab to Mass Customization*,
PoliMI SpringerBriefs, https://doi.org/10.1007/978-3-031-73327-7_1

of instructions, nearly like a precursor of the Ikea model, but for full houses. The system allowed flexibility and affordability to respond to the requests of a large number of blue-collar workers.

Le Corbusier at the start of the Twenty century was a fervent advocate of prefabrication, as elaborated in his "Mass-produced Houses" essay [1]. His thoughts gave a solid foundation to the International Style, which made large use of prefabricated components for multistorey buildings.

The development of the steel industry for buildings provided a big push for prefabrication. Among the early experimentations, the Dymaxion House by Buckminster Fuller is certainly worth a mention. Presented for the first time in 1929, during the American Great Depression, the Dymaxion House was granted its name by Fuller's main ideas of "dynamic, maximum, tension". It was a hexagonal/circular house, in which the structural steel and aluminum components were held together in tension suspension from a central pole. The house was perfect for a family of four, with two bedrooms and two bathrooms, and already anticipated the current ideas of circular and resilient buildings, as it was designed to be easily assembled, disassembled, and reassembled, and it could withstand fire, earthquakes and high winds. Despite the obtained attention and the futuristic approach, it took twenty years for this proposal to be built. It was indeed, only during the war, that the Beech Aircraft Company chose the Fuller project for housing their workforce, given its low cost, easiness to be maintained, and capacity to withstand tornadoes.

Frank Lord Wright also was interested in solving the problem of building affordable houses for Americans, and with his Usonian House, he approached prefabrication, by developing a grid system that defined modular dimensions for wooden houses. The idea of the grid system is a precursor of the "Platform approach", that will be discussed in Chap. 3. The Usonian grid system would allow the use of standard wooden components, and, consequently, reduced costs. The components assembled in a flexible grid scheme would provide a level of customization to make each design unique, to some extent. The Usonian houses offered well-built, cozy, and—above all—affordable dwellings. About two dozen Usonian houses were built between the thirties and forties.

Industrial production for building systems has been since then continuing to evolve. Another renowned case is Seidler's design commissioned by the Royal Australian Institute of Architects Convention, which was constructed starting from prefabricated components developed with available local materials and was characterized by great flexibility of composition. The system was influenced by Gropius Bauhaus's principles and could be erected by four men in a day. Research in developing prefabricated systems that could break the monotony of mass-produced housing, which was mostly common up to that moment, was also the intent of Jean Prouvè. He developed a system, based on jointed steel structures that could be erected without scaffolding, and could give form to fourteen variations applicable to a variety of building typologies, including housing and hospitals, demonstrating, this way, the possibility of producing on large scales.

Already mid-twenty century, obtaining a mortgage for prefabricated houses was a thorny issue. But the Housing Act of 1949 recognized the American housing crisis and gave a further push to the development of mass-produced systems. Levitt was an important pioneer at that time, bringing the factory to the site, developing entire housing communities, and ending up building 6000 houses in Levittown, for veterans coming back from wars and their families. Levitt adopted a large-scale manufacturing approach, based on the procurement of all the materials, modular components, and supplies on-site, and the identification of 27 building processes, that were performed in sequence by builders, roofers, and plumbers across multiple houses, which were erected contemporarily. In this way, he managed to build a house in 17 min. keeping drastically down the costs, and allowing the buyers to access financing. Moreover, he set up a standard for the housing systems that followed in terms of space and function articulation and included also copper radiator systems, aluminum Venetian blinds, and insulated glasses. Levittown initiated the American mass-produced suburbia.

Quality of life was brought to a further level by Eichler. He like Levitt started to develop full communities of housing, but incorporating also green spaces, open plans, glass walls, two bathrooms, and an inner opened atrium. His architecture emphasized the relationship between the inside and the outdoor environment, shaping, this way a new lifestyle for young buyers.

In 1945, the first historical large call for Case studies development of mass-produced housing was launched by John Entenza, and titled "Case Study Houses". This saw thirty-six prototypes built over twenty years, that were designed by master architects such as Richard Neutra, Charles and Ray Eames, among others. Many of them were steel prefab houses, with steelwork left exposed, to celebrate the honest materiality of the system, in line with the modernist ideals of true expression of the materials. Many of these houses used prefabricated components that were carefully assembled and designed to allow the natural flow of light and air, as well as to dialogue with the natural landscape. In a sense, they can be considered the triumph of the idea of mass-manufacturing architectural masterpieces. They were aimed to make tangible the T Fordian approach to mathematical rigor for comfortable mass-manufactured housing.

However, regrettably, what followed after these masterpieces were mediocre mobile-home parks, fuelled by manufacturing companies that were interested more in capital and financing rather than the design of comfortable houses. After the Wars the request of building for many to be housed in a short time allowed the construction of too many low-quality housing and construction. This undermined prefabrication, to the point, that many industries started to label their product differently, in the hope of distancing themselves from the growing general idea that equated prefabrication to low-quality, boring, and same-looking buildings.

In response to that general view, Techbuilt was founded by Carl Koch, who developed a system of components that could be assembled by owners according to their own needs and tastes. This advanced the idea of prefabrication methods that can lead to mass customization. It was the fortune for that system and for others that looked like typical colonial or vernacular buildings while having an industrial process at their heart [2].

A very unique example of an industrialised system, which was developed for tropical climates and to be flown to African colonies, is "La Maison Tropicale" designed by Jean Prouvé in the 1950s. Designed to house French colonizers, was completely made of aluminium sheets, steel components, and wooden screens. It anticipated its time for the well-articulated adoption of natural ventilation strategies for hot climates. It used heat from the house's roof to convey fresh air within openings in the walls and ceiling. To reduce the heat gains, double-skins insulated walls and sliding doors with openable holes were used as a façade system, while the verandas were shadowed with adjustable parasols. Only a few of them were built in Niger and Congo, as their cost resulted in prohibitive for many. However, La Maison Tropicale design is still considered today of great interest, so that it was recently rebuilt in a London exhibition in 2008, as a tribute to the French architect, and as a demonstrator of how to build new home quickly and well.

Around 1960, again, many architects started wanting to develop housing that could provide a proper house to many, as the basis for providing equal rights to communities and addressing the energy crisis. The Zip-Up house was designed in 1968 by Sir Richard Rogers, in response to the competition "House of Today" for the "Ideal Home exhibition", and it was a self-supporting aluminum tube, made of curved aluminum sandwich panels, with a thickness of 20 cm, that thanks to their thermal properties needed a very low (for the time) heating system. The modular system could have been, theoretically, infinitely extended. The client could have bought any quantity of the self-supporting panel, assembled them with some neoprene, and by using a tool similar to a tin opener, made the space for windows and doors [3]. The house would stand on a steel frame with concentric bracing, having adjustable columns to be located in any setting, with minimal impact on the site. This supporting system was also modular and could have been infinitely extended for the house. The house system as well as the furniture and services were conceived to be fully flexible, with a kitchen on wheels, removable internal walls, and modular services. Rogers with his studio has continued the exploration of modular systems throughout his career, with proposals for more light volumetric units, as in the case of the Y: Cube which was developed in collaboration with the Sheffield Insulation Group, manufactured off-site, with high thermal insulation performance, and fully furnished, made through precise laser cuts, to be perfectly fit together in the final on-site assemblage. Firstly adopted by the YMCA of London, for those leaving the YMCA for longer-term accommodations, and then developed into larger units for the PLACE/Ladywell, where twenty-four family houses are assembled together, each of them starting from two volumetric units, of which those on the ground floor have higher ceilings and are devoted to community functions, as for shops, offices, and café.

The research toward modular and energy-efficient systems continued also with the Advanced Technology House by Kappe and commissioned by NASA, which developed a volumetric unit that could be assembled horizontally and vertically and could adapt to various climates. The unit included glass skins that could change from clear to opaque according to solar conditions and also had power units that could be disassembled and transported for future uses.

The other line of modular systems that have been developed since the sixties is the "capsule architecture". Nakagin Tower in Tokyo was the first built example and aimed to create autonomous volumetric spaces for families that could transform to accommodate new requirements, such as transforming a single flat into a two-bedroom apartment, for a growing family.

In summary, over the last century (Fig. 1.1), prefabrication has evolved many of the fundamental principles that we see today and will be discussed in the following chapters, enquiring about their capacity to deliver products able to respond to the current high housing demand, while providing the required performance in terms of building safety, thermal comfort, energy efficiency and aesthetics, throughout all their life-cycle.

1.2 The Case of the Heavy Mass Production in Between the World Wars

Before investigating the current Modern Methods of Construction, it is necessary to cite, the heavy prefabrication that has been developed on a large scale across all of Europe throughout and after the World Wars. For those living in European cities, multistorey precast concrete blocks are still the main testimony of prefabrication used for large social housing estates, and, hence, are the first term of reference.

Among precast modular concrete buildings, Large-Panel Structure (LPS) buildings are realized by assembling large structural components (floor slabs and wall panels) manufactured in a factory and assembled on-site in a short time and at a low cost [4]. This building typology has largely been used after the Second World War to house a growing population. The first multi-storey residential buildings were built at the end of the Fifties in the Soviet Union, and since then 40% of construction in socialist countries was made by Large Panel Systems, and a strong use has also been made in many capitalist countries.

In the UK, in between the wars one million dwellers were built and an extra four million dwellers were built after the Second World War [5]. They aimed to house the refugees, and clear the slums, that characterized many British cities. Their construction in the UK was strongly supported by Churchill, and aimed to materialize the modernist ideals of mixed developments of towers for residential housing, erected above car infrastructure, and within green areas. They also aimed to become the symbol of modernity, with the integration of electricity, new domestic tools such as television, laundry, ironing, and more household tools, that were for the first time in history entering the middle-class housing market. Hence, in some terms, the use of industrialized systems for housing construction was seen as the perfect interpretation and materialization of the "House as a machine" idea, strongly elaborated by Le Corbusier and all the modernist architects.

In terms of technology, pre-cast panel systems were usually used in two main design schemes, depending on load-bearing structure: (1) Frame Panel—all the base loads are borne by the building's frame, and the panels are used as enclosure elements; (2) Large-Panel (frameless) System (LPS)—the structure is based on the assembly of large panel for walls and slabs that perform the load-bearing and enclosing functions

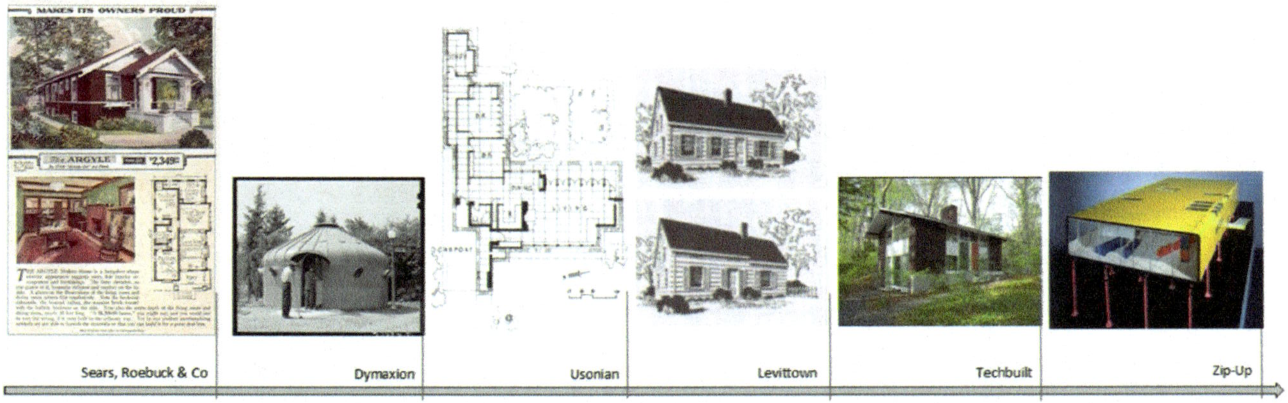

Fig. 1.1 Evolution of prefabricated houses showing from the left side: the kits commercialized by Sears, Roebuck & Co., Dymaxion House by Fuller, Usonian House typical plan by Frank Lord Wright, the house for Levittown, the Techbuilt by Carl Koch, and the Zip up by Richard Rogers

simultaneously. Large estates of the second typology are still present in the European residential building stock [6]. LPS are gravity structures, comprising precast reinforced concrete floor and roof components, spanning onto one-storey high structural precast concrete wall panels (Fig. 1.2). Components are connected by various forms of dry or wet joints made on-site. Vertical loads are carried to the ground through the structural wall panels which also provide stability against lateral loads. Thus, the joints are the most vulnerable part of a LPS. When properly joined together, horizontal elements act as a diaphragm that transfers lateral load to the wall panels. Depending on the wall layout there are three configurations of LPS buildings, namely cross-wall (flank–wall if the wall is external and located at the end of the buildings), spine-wall, and two-way systems if the main walls that resist gravity and lateral loads are oriented along the short or the long or both direction of the building, respectively. Panel connections represent the key structural components in these systems. In particular steel reinforcement in the horizontal wall/floor joint has no significant role in the primary structural function of carrying vertical loads to the ground in normal loading conditions. However, in case of accidental load, the horizontal joint is of primary importance to mobilize continuity forces to avoid local failure [7]. If the accidental loads are of large magnitude, local damage could be caused to load-bearing components and/or joints. In that case, an alternative load path to maintain the overall stability of LPS blocks needs to be ensured. LPS buildings are inherently more vulnerable because of the difficulties in providing continuity and ductility in such a system. The vulnerability of LPS construction under accidental loads gained interest in 1968 after the partial collapse of the southeast corner of Ronan Point's 22-storey apartment block in London (UK), due to a gas explosion.

Fig. 1.2 Large panel systems in UK

That event, together with the turn of the utopian ideal in dystopian social reality, with many of the UK tower block neighbourhoods suffering from social unrest, brought to the destruction, starting from 1993, of some of the largest developments and created a historical precedent that questioned the application of large-scale precast concrete systems. These towers, specifically, are still of great interest today, as they still represent a large built asset, that should be retrofitted to satisfy the current energy and comfort requirements. More will be discussed in the final chapters of this book.

1.3 Learning from Past Failures

Over time, the construction sector has witnessed periods of failure of prefabricated systems. When prefabrication has been associated with infinite rows of poorly constructed, charmless housing or warehouses, they have been regarded as failing for being cheap, ugly systems built with substandard solutions. The economic benefits of developing systems to be manufactured in a factory through the use of standardized components, which have drastically reduced the erection time, and have brought them to be chosen by many builders and developers, have been paid in terms of aesthetics, comfort, and quality. Examples of this approach can be US cottages.

However, exceptions of value coupled to periods of economic optimism and technological advancements, have allowed over time prefabricated systems to resurge and be used for new construction as well as for retrofitting existing buildings, as we are currently witnessing. Therefore, to sustain the growing impetus in prefabrication, a critical understanding of two major failures needs to be considered, and they are the Ronan Point event in Newhan, London (1968) and the Grenfell Tower fire (2017) (Fig. 1.3).

These events require attention, not to stigmatize all the industrialized buildings or building components, but instead, to understand, what potential weaknesses contemporary and future systems need to overcome.

The collapse of a twenty-two storeys tower block in Newhan, London (1968), due to an internal gas explosion, demanded attention to workmanship and details. Specifically, the vast investigation campaign developed after the collapse of the Londoner Tower, demonstrated that the large panel building typologies can be susceptible to accidental loads, due to the low robustness of the connecting systems between walls and floor panels. Accidental loads are described as events that result from an unanticipated extreme cause, such as a car bomb attack, gas explosion, impact, or mistakes made during the design or construction stages. These events can cause localized damage, partial collapse, or full collapse if there is insufficient continuity, ductility, or redundancy to stop the damage from spreading [8]. Hence, the event demonstrated that for that specific building typology, the correct executions of the connections, made by interconnected steel rebars between wall panels and floor slabs steel hook, and sealed with compacted concrete were essential to the safety of the overall building. Hence, one of the main lessons learned from the tragic collapse is that if

Fig. 1.3 Ronan point collapse and Grenfell tower

connections are viewed as merely site-specific features rather than as an integral and necessary component of the structural system, then prefabricated constructions will not meet safety and reliability requirements. The efficient planning and development of these structures is made possible by connections that are adequate for all service, environmental, and ultimate load scenarios, and that are properly executed on-site, during the assembly of the prefabricated panels [6]. Moreover, these prefabricated tower blocks, built between 1950 and 1970 are reaching the limit of their service life, and in some cases, may have already surpassed it, which increases the likelihood of serviceability problems. Therefore, these buildings require today to be retrofitted to meet contemporary safety and energy standards [4]. Accordingly, the requirement for retrofitting this type of old building becomes a top European challenge.

The Grenfell Tower fire is instead a tragic event, that happened to a block of flats in North Kensington (London) and was originated with a fire that started in a fourth-floor apartment and quickly spread around the exterior of the 24-story building, engulfing all of the residential floors in smoke and flames. Only a few years prior, the tower block had gone through an energy retrofit. During that time, the energy efficiency of the building was improved by applying a so-called "façade-first approach", which entailed the reduction of energy consumption through the improvement of the energy performances of the façade, by the addition of a ventilated double skin, made up of prefabricated components. The tragic 2017 fire accident showed that the precision and economy of the solution, as well as, the delivery short time, were at the expense of using insulation material, which was found to be flammable material, which burnt for roughly 60 h and claimed 72 lives. Following the incident, the majority of towers in the UK underwent verification, and it was discovered that many of them were covered in materials that were comparable, which amply demonstrated that the case was not isolated and that the use of flammable material in prefabricated systems for façade retrofit was a systematic issue. The following Grenfell Tower inquiry has required a revision of the fire assessment for new prefabricated systems, as well as the requirement of integrating fire design principles within the academic curriculum of civil engineering, architecture, and architectural engineering in the UK.

Each of the cited events has resulted in the requirement for new or updated regulatory frameworks for progressive collapse prevention (as dictated by the Ronan Point collapse), and fire regulations (as following the Grenfell fire event). Moreover, both subjects have entered the academic curriculum, to develop professionals with the necessary background knowledge and awareness, which are "sine qua non" requirements for the development of safe and enduring prefabricated system.

1.4 Why Modern Methods of Construction Now?

In 2005, the UK government started to proclaim a large housing crisis, due to an outdated housing stock and the necessity to deliver thousands of affordable housing. The government was interested in developing new building techniques and processes to accelerate building provision. A report by the National Audit Office defined "Modern Methods of Construction as a process to produce more, better quality homes in less time" [9]. The report, building on the results of four workshops with major UK stakeholders, aimed to determine the best way to get value when employing MMC. The indications were cautions in indicating the benefits of adopting MMC, demonstrating the opportunity offered in terms of speed to deliver the required houses, and potential quality of achievable perforances, but it also highlighted the risks of investments, and hence the necessity of appropriate risk management and liaising with planning authorities. About ten years later, in 2016, with an increasing surge in housing needs, UK construction sector was shaken by "The Farmer Review of the Construction Sector", also known as "Modernise or Die" [10]. The published detailed review diagnosed a construction sector near failure due to industrial trends and societal transformations. Cost inflations, already in 2016, i.e., just after Brexit and before any other major shock, such as COVID-19 and the Ukrainian War, were making construction nearly unaffordable, with initial signs of major projects having important delays, and/or delivering not at the necessary quality level, and/or going bust. In particular, the report pointed out the major impact that construction was starting to see due to workforce demographics, and skill shortage. While, indeed, until that point UK was solving the skill shortage issue with immigrants [11], the soon-entering force of Brexit, started making that difficult, to the point that Farmer estimated a 20–25% decline in the skilled workforce within a decade. This scenario depicted Britain as incapable of delivering the required infrastructure, and housing, as well as the expected national GDP.

Moreover, the technological changes in society, with the digitalization of information and processes, would bring a new generation of workers closer to a digital world rather than to manual craftsmanship, with a consequent reduction of attractiveness to construction works. He also highlighted the importance of changing the tenure approach, and that contemporary society may need more of a diversified tenure model. However, house tenure is outside the scope of this book.

The report defined a decalogue of recommendations that incited the coordinated collaboration between industry, governments, and clients, supported by academic expertise to develop and deliver an innovation program, that could shape a diverse demand for construction in general, and specifically for residential buildings. It also pointed to the necessity to reform the Construction Industry Training Board (CITB), to deliver knowledge and skills aligned with digital expertise. The document called for the government to act to incentivize the adoption of pre-manufactured solutions with a variety of measures, from policy, to direct commissioning of pre-manufactured buildings, or promoting prefabrication through a list of registered providers. All the incentives, upskills of the workers, research and development for industrial innovation, and the changing of the training boards, should be done with a long-term vision, and therefore a national pipeline of demand for housing was advocated.

Indeed, companies that were investing in innovation, seemed to be having trouble bringing new ideas and products to market on any kind of large scale. This was the result of ingrained risk perceptions among building control inspectors, clients, advisers and designers, insurers, and investors, as well as the larger supply chain. Adoption appeared to be hampered by the requirement for a strong, if not absolute, benefits argument. Instead, many novel methods of construction design and processes are often automatically regarded as high risk due to negative attitudes. The multiple financial collapses of companies that made significant investments in various strategies that would have increased construction productivity and predictability but were never successful do not assist with this.

The document also provided a series of critical thoughts and pieces of evidence, that the industry will miss the best chance to increase productivity and counteract workforce loss if it does not fully embrace the current speed and nature of technology innovation and change in society. To demonstrate, instead, the potential benefit of a changed approach to manufacturing houses, Tokyo was mentioned as a virtuous example, capable of producing almost as many houses annually as the UK can on a national level. This was primarily because single-family homes are delivered utilizing a model, that capitalizes on the mass market's cultural acceptance of pre-manufactured modular housing. This in turn provided empirical evidence for the possibility of successfully offsetting labour shortages through delivery method changes in the housing sector. However, it is also clear that a cultural acceptance factor plays an important role in the success of mass-modular housing in Japan.

In 2017, the UK Government with the white paper "Fixing our Broken Market" [12] declared to promote cutting-edge techniques for housing construction. Through partnerships with smaller and medium-sized builders and contractors in the Accelerated Construction program, as well as assistance in helping them obtain the loan financing they require, the government aimed to promote a wider diversity of homebuilders. In exchange, the government required developers to focus on design and quality, build more homes, interact with communities to build confidence in the MMC developments, and build quickly.

1.5 The Newly Defined MMC Categories

The aging of housing and real estate, coupled with population growth has given new impetus to the development of prefabricated methods [13]. In particular, in the UK, after the publication of the white paper "Fixing our broken housing market", in 2017, a cross-industry working group was established to look at the variety of construction technologies and the new technological advancements brought by most recent manufacturing processes, and defined a new framework to categorize the wide range of cutting-edge building methods that are being used in the residential market, and are expected to expand in the near future. The group, which included representatives from Homes England, the UK National Home Building Council (NHBC), and the Royal Institute of Chartered Surveyors (RICS), defined seven categories of MMC [14].

Site-based materials, process innovation, and premanufacturing in all their forms (i.e. manufacturing in remote, near-site and pop-up factories) are covered under the MMC categories. The MMC framework was developed to increase stakeholder knowledge and education of MMC and to facilitate improved access to mortgage financing, insurance, and assurance.

Category 1 defines the "3D primary structural systems", which corresponds to the most common idea of volumetric units, that certainly include some structural elements and can also contain internal and external fittings, services, and/or envelope components (external cladding or roofing). Volumetric units also include pods, as for kitchen and bathroom units. The volumetric units can be assembled on-site in pure volumetric buildings, or in hybrid configurations, by combining volumetric units with traditional on-site manufactured components or combining them with other pre-assembled systems.

Category 2 defines "2D primary structural systems", and includes flat panels for floors, roofs, and walls, that are fabricated in a controlled factory environment, and can either include only the skeleton structure or also contain, in the so-called "closed panels", the insulations, doors, windows, finishing, and cladding. The panels are to be assembled on-site, generally through mechanical or chemical connections, and can be combined with 3D systems or other traditional on-site technologies.

Category 3 defines "Non systemized structural components" as pre-manufactured linear structural members such as beams and columns, or pre-manufactured structural walls or slabs, which can be made of a variety of materials, from hot-rolled, or cold-formed steel, wood or precast concrete, or any new future construction materials. In other words, this category consists of what previously were called Stick-built construction systems (Iuorio 2009), prefabricated frames, preassembled trusses, and piles.

Category 4 includes "Additive Manufacturing" for both structural and non-structural components. This category is, with no doubt, one of the most innovative, as it recognizes the role and the impact that digital design and manufacturing are starting to have, and acknowledges that these new design and manufacturing processes can, after years of intense research and development, is becoming mainstream in construction. This can include 3D printing of concrete buildings, 3D printing of ceramic or metal components, as well as fiber wiring. All processes, that can speed up construction, and reduce the amount of material used through geometrical and performance optimization, consequently reducing waste and increasing customization.

Category 5 defines "Non-structural assemblies and sub-assemblies", and includes all those components that have been pre-manufactured but have no structural capacity. These can be non-structural walls, prefabricated roofing components, façade systems, kitchen or bathroom pods, mechanical and electrical (M&E) units, and many other industrial assemblies that can be used in construction, as derived from a factory process. These systems can be combined with any of the previous categories for building construction.

Category 6 includes "Traditional building product-led site labor reduction/productivity improvements", which are all the traditional products that are pre-formed and cut in a factory and used to build. They include large-form wall products for internal and external walls, as well as pre-sized and pre-cut components.

Category 7 includes "Site process led labor reduction/productivity improvements", which are all those processes that can improve productivity and reduce on-site labor in construction, through the use of lean processes, digital and physical worked augmentation (such as augmented and virtual reality for construction workers), exoskeletons, and wearables, or robotics and drones. This category, in line with category 4 opens up the possibility of considering as MMC also those construction systems having processes that embrace new automated and or human–robot collaborative procedures. This category is flourishing and has seen many applications in infrastructure and building assessments, such as laser scanning through drones for crack or building pathologies assessment, drones or robots for plastering, high-speed panelizing systems for screwing for build-up wall or floor panels (see also Chap. 3), collaborative robots for lean production, or building in harsh environments, and many more.

Categories 6 and 7 can be used in conjunction with categories 1 to 5, to further reduce waste, increase productivity, and reduce the amount of supervision through the increase of digital planning and the use of augmentation. Figure 1.4 summarizes the presented categories.

1.6 Conclusion

This chapter aimed to recall the main historical evolution of prefabrication looking primarily at Europe and USA. The historical evolution specifically aimed to identify the development of diverse systems and manufacturing approaches, from the

Fig. 1.4 The seven categories of MMC

catalogue of industrialized elements and components to near-site factories to heavy prefab, to retrace the gems of what we are seeing today in currently under-development systems. It also discussed the "elephant in the room", by investigating the failures, with the intent to set free to look at current systems without prejudices.

Finally, it traces the most recent governmental attempts that in UK are pushing to the resurrection of industrialised systems under the new umbrella of Modern Methods of Construction. The newly defined categories of MMC are embracing the changes and diversification of manufacturing processes, reflecting recent breakthroughs in digital technology that aim to reduce manufacturing costs. The new categories are providing a framework for a large variety of small and medium enterprises, large industries and spin-offs to develop systems for small and large-scale applications. This is incentivizing both research and development in the field as well as real-case applications.

References

1. Le Corbusier (1985) Towards a new architecture. Dover Publication
2. Arieff A, Burkhart B (2002) PREFAB, Gribbs Smith Publisher. ISBN 1-58685-132-2
3. Rogers R (2017) Un posto per tutti. Vita, architettura e società giusta. Johan & Levi Editore, ISBN 978-88-6010-211-9
4. Iuorio O (2018) Energy retrofit of tower blocks in UK: making the case for an integrated approach. Techne: J Tech Arch Environ S1:45–48
5. Muthesius S, Glendinning M (2017) Towers for the welfare state. An architectural history of British multi-storey housing 1945–1970
6. Romano E, Iuorio O, Nikitas N, Negro P (2018) A review of retrofit strategies for Large Panel System buildings. In: IALCCE 2018: The sixth international symposium on life-cycle civil engineering life cycle analysis and assessment in civil engineering: towards an integrated vision:

proceedings of the sixth international symposium on life-cycle civil engineering (IALCCE 2018). Taylor & Francis (CRC Press), pp 3023–3030
7. Matthews S, Reeves B (2012) Handbook for the structural assessment of large panel system (LPS) dwelling blocks for accidental loading. BRE Press. 978-1-84806-200-9
8. Fib (2012) Design of precast concrete structures against accidental actions. Bulletin 63. International Federation for Structural Concrete, Lausanne, Switzerland
9. National Audit Office (2005) Using modern methods of construction to build homes more quickly and efficiently, Report, file:///C:/Users/iuori/OneDrive%20-%20Politecnico%20di%20Milano/Papers/BOOK/Chapters/Chapter%201/biblio/mmc%202005_UK.pdf
10. Farmer M (2016) The farmer review of the UK construction labor model. Modernise or Die. Published by The Construction Leadership Council (CLC)
11. Rolfe H, Hudson-Sharp N (2016) The impact of free movement in labour market: case studies of hospitality, food processing and construction. National Institute of Economic and Social Research (NIESR)
12. Department for Community and Local Government (2017) Fixing our broken market. ISBN 9781474137959
13. Green SD (2022) Modern methods of construction: reflections on the current research agenda. Build Cities 3(1):653–662. https://doi.org/10.5334/bc.265
14. Independent Report (2019) Modern methods of construction working group: developing a definition framework. https://www.gov.uk/government/publications/modern-methods-of-construction-working-group-developing-a-definition-framework. Accessed on 30 Jan 2024

Chapter 2
Additive Manufacturing and Automation in Construction

Abstract The two most disruptive innovations entering the construction sector in the last decades are additive manufacturing (AM) and automation in construction, which are indicated in the new categories of MMC as categories 4 and 7. This chapter aims to frame the development in these fields by adopting an approach that goes from the material to the construction process scale, highlighting the importance to adopt fully integrated design and manufacturing processes. Material advancements through AM in clay, earth-based, steel, and bio-composites are discussed to reflect about the mainstreaming of AM processes to create high-performance low-carbon building components. It presents a generalized four streams framework for robotics in construction, that can be used for understanding and analyzing advancements in a field in continuous transformation. This chapter discusses immersive reality as complementary to automation to explore multiple design opportunities and to monitor construction and maintenance phases. In an ideal world where robotic manufacturing and assembly are considered from the inception of building design, the domains of robotic, advanced manufacturing, architecture and building construction can influence each other opening up new formal and technical languages, rooted in traditional sciences.

Keywords WAAM · Bio-composites · Low-carbon materials · Robotics · Immersive reality

2.1 Additive Manufacturing

Additive manufacturing (AM), also referred to as "3D printing," is a cutting-edge technology that has the power to transform the building sector, address housing shortages, and lower greenhouse gas emissions. AM has been largely advanced in the last decades thanks to the opportunities it can offer in going beyond the technical and process limitations of traditional construction methods to enable the development of materials, geometries, and systems tailored to specific needs. AM terminology has

O. Iuorio, *From Mass Prefab to Mass Customization*,
PoliMI SpringerBriefs, https://doi.org/10.1007/978-3-031-73327-7_2

been standardized through the ISO/ASTM 52900, which provides general descriptions of the most developed AM processes while establishing a terminology that can allow continuous development of AM across various fields. AM is continuously advanced as it can bring environmental and social sustainability benefits that include enhanced resource efficiency through optimization of material use (see Sect. 2.1.1) and customized building components (see Sect. 2.1.2), while having the potential to enable higher productivity in the construction industry (see Sect. 2.1.3).

2.1.1 3D Printing for Material Optimization

Under the AM umbrella are clustered all those processes that allow the making of objects from 3D model data [1]. The first developed AM process for construction, which is still largely used for fast prototyping, is the fused-deposition modeling (FDM) technique. FDM is a material extrusion (ME) technique, one of the seven AM categories defined by ASTM F2792 [2]. This technique uses thermoplastic material, which has low melting temperature, such as polylactic acid (PLA) or acrylonitrile butadiene styrene (ABS), to melt it from a solid to semi-liquid state, to allow it to flow through a heated extrusion nozzle and deposit it layer by layer onto a printing bed (Fig. 2.1). The material and mechanical quality of the final object are affected by the layer thickness, width, temperature, printing orientation, and environment. The objects printed through fused deposition have usually anisotropic properties, with weaker tensile and yielding strengths along the layer-building. The anisotropic properties of these materials limit the application for structural systems, while the rapidity of production, and the easy interface with many computer-aided-design (CAD) and modeling software make it excellent for fast prototyping.

Fig. 2.1 Fused deposition processes in PLA (on the left) and in metal (WAAM, on the right)

The fused deposition process is also used with metal wire and is known as Wire and Arc Additive Manufacturing (WAAM). This process involves melting a metallic wire and extruding it onto a metal substrate that serves as a bed (Fig. 2.1). The WAAM process was initially patented in 1925 [3] for metallic ornaments, but only in the last decades has started to be developed for construction and infrastructure applications. In construction, specifically, its implementation is driven by the possibility to produce intricate and complex geometries, that cannot be achieved through traditional steel systems. The possibility to develop with WAAM tailored geometries for beams, joints, and trusses, can allow the development of higher-performance components, while reducing steel material usage, with consequent weight reduction of the overall construction system, which can bring benefits to the overall substructures. Moreover, given the relatively high embodied carbon of traditional steel, reducing the amount of material utilization could lead (in theory) to lowering the environmental footprint. Of course, as with any new technology, there are multiple variables to be considered, and they are under the investigative lens of researchers. In the WAAM continuous process, the metallic wire is made to melt through the energy of the wire arc, and the metal is deposited layer-by-layer on the metal substrate, according to a CAD-based model. In this process, metals' weldability is crucial since the procedure is akin to arc welding in certain aspects, and, indeed, three welding technologies have been developed for WAAM [4]. Thus far, this technology has been developed to be used with wires made of steel, titanium alloys, aluminum alloys, and alloys based on nickel. For each WAAM application, choosing appropriate process parameters is essential to obtain a neat shape, and produce sound components. To this aim, the main parameters to control are wire selection, slicing direction (which can be uni-directional or multi-directional), the tool path (as this can influence crack formation), the welding process and its related parameters (such as wire feed rate and travel speed), and shielding gas composition and flow rate. It is important to note that elevated heat input levels in WAAM can limit high-quality production and lead to various defects and challenges in materials processing such as high residual stresses, porosity, delamination, and cracking. These challenges need to be addressed to further develop this technology and make it suitable for more applications in the construction industry. WAAM is gaining popularity for producing large-scale components including structural elements because of its huge build size, high deposition rate, and affordability. In contrast to other steel AM processes, such as directed energy deposition (DED) techniques and laser metal deposition (LMD), which can produce more precise part finish at the penalty of higher prices, reduced efficiency, and smaller build volumes, WAAM compromises on dimensional accuracy and surface quality.

A kind of FDM can also be achieved for clay and stoneware materials. For them, AM extrusion has enabled the production of ceramic components with specific, intricate design requirements that are difficult to achieve with traditional methods. The process can involve Controlled Numerically Cutting (CNC) machines with a three-axis movement system, driven by a stepper motor, which pushes a plastic ceramic paste through a die. The material flow, printing velocity, and air pressure are controlled both computationally and manually, adjusted according to the paste's viscosity and the specific design requirements, with print speeds ranging from 20 to

100 mm/s. For AM clay and ceramic materials, one of the key parameters to control is the water content, as it influences on one side the extrudability of the paste, and on the other side the plastic deformation that can occur at the printed components, after being manufactured.

AM has also been largely developed for concrete printing, and it is starting to find application in the production of both building components and full housing, as reported in the following sections. 3D-printable cementitious materials have different properties from regular concrete, including rheology, printability, and compressive strength. To determine the critical design parameters that ultimately determine the performance of 3D-printed concrete, some authors suggest adopting a multi-level material design (MMD) approach [5], taking into account the three primary fabrication stages, which are the design of the mixture, the printing procedure, and the composite structure. The development of 3D concrete printing (3DCP) applications for non-conventional building and architectural components had some initial applications developed by teams at Loughborough University. That group created computer-controlled 3D printers that deposit high-performance concrete layers precisely, creating intricate structural elements that are impossible to create with traditional methods, like curved cladding panels having also intricate textures and architectural features. The method has also been developed, in partnership with Skanska, to incorporate increasingly complicated building services infrastructure early on in the design and construction process. Concrete 3D printing today is fast developing worldwide with some major applications in Asian countries.

Advanced Materials present promising benefits that can lower material consumption and environmental impact. A recent study by Bianchi et al. [6] aimed to evaluate the environmental impacts and thermal characteristics of different geometric configurations and infill patterns in 3DCP systems made for components of the building envelope, and which have been produced with three different AM technologies: FDM, small gantry crane-based 3D printers, and large gantry crane-based 3D printers. The study found that 3D printing technologies offer low energy consumption and sustainable alternatives to traditional structures. However, their impacts depend on the configuration and internal infill, with thinner wall thickness improving envelope sustainability. Of course, the latter continue to be more challenging for large-scale applications and therefore require further development.

2.1.2 3D Printing for Customized Building Components

AM has progressed in developing prefabricated building components at diverse scales. We have seen it employed off-site for making architectural clay bricks, structural components like columns, joints, and slabs, and more multifunctional components such as wall panels, always with the intent to advance customization of the components, starting from their geometry, to improve aesthetic quality, while controlling structural, constructional, thermal and acoustic performance.

AM processes for stone bricks have been developed by the group of Cruz and Figureido, and their team in the Advanced Ceramics R&D Lab since 2016, to produce free-form bricks for building facades that could integrate customized requirements. Stoneware is a ceramic clay based material, which is fired at high temperature to reach superior mechanical properties and low porosity, in comparison to traditional clay. The fire process makes stoneware water-resistant and durable. In their initial work, the Wave Wall the S-Brick Wall and the Hive Wall prototypes, they investigated the possibility of producing with 3D printed bricks vertical wall curvatures, double curvature sinusoidal surfaces, and non-uniform truncated pyramids for ventilated walls [7]. In particular, with the S-Bricks (Fig. 2.2 left) they investigated s-shaped configurations, for the overall bricks geometry and their inner structure. This allows for layering in two directions, creating components with varying designs and functions. When the bricks are assembled in the extrusion direction, they produce an opaque wall featuring a corrugated pattern, providing both aesthetic appeal and structural integrity. Conversely, when the bricks are laid flat, they form a ventilated sunscreen wall. The S-shaped curves play a crucial role in this configuration, as they facilitate a secure and stable connection between the different layers of bricks, enhancing both the structural performance and the ventilation capabilities of the wall. This adaptability in brick assembly allows for innovative architectural applications, ranging from solid, visually striking surfaces to functional, breathable facades. At the basis of their work, there is also a systematic classification study of a range of stoneware pastes for AM applications, that included the definition of the ideal moisture content to be assured for obtaining an acceptable plasticity, while warranting a proper adhesion between extruded lines and limiting potential plastic deformation, that could happen after the shaping process.

AM technologies for column applications have caught the attention of multiple research groups, and numerous results in clay, bio-composites, concrete, and steel

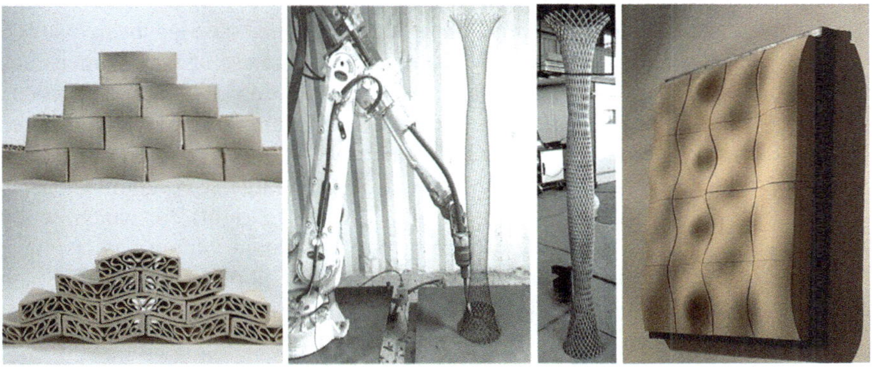

Fig. 2.2 AM building components: (left) curvilinear stone bricks with advanced thermal performances by (Cruz and Figueiredo); (center) steel diagrid column in WAAM by Laghi et al. [8], (right) biocomposite 3D printed wall panel system by BioMat, as displayed at Biennale of Architecture 2023 in Venice

would be worth to be noted. Among them, this section focuses on the stainless steel diagrid column proposed by Laghi research group at the University of Bologna. The prototype was developed through a dot-by-dot WAAM technology. The dot-by-dot technology differentiates itself by the continuous WAAM technology, as the material is deposited through sequential points. Hence, the result of a dot-by-dot process is a rod-like element, developed along a longitudinal axis, which can be straight or curvilinear. The diagrid application, as shown in Fig. 2.2 center, studied the development along curvilinear axis, where the latter were defined based on a computational optimization design process. The overall column shape and the topological design were defined with the objective of minimizing weight and optimizing utilization factor while considering historical precedents and manufacturing constraints. These objective functions were set in Grasshopper and through Galapagos a population of solutions was generated, until the final design of a 2-m high non-uniform cylindrical column was defined. The CAD design was then implemented in a robotic path planning, as the prototype was printed through a robotic arm, which was equipped with off-the-shelf welding tools. The obtained column proves the capability of WAAM technology also at large scale and for the development of thin components. Their study [8], also indicates the requirement of improving the printing quality, reducing imperfections, and developing fully integrated design and manufacturing processes.

Looking at the sphere of alternative low-carbon materials, an interesting and promising perspective has been opened up by BioMat@Stuttgart and BioMat@Copenaghen, which is advancing the development of construction components, including façade systems, partition walls, and columns, utilizing 3D-printed biobased materials. These innovative biocomposites consist of a fiber and a binder, with at least one constituent derived from biomass. By integrating natural fibers and biobased binders, these materials offer sustainable alternatives to traditional construction materials. The use of 3D printing technology allows for precise fabrication, optimizing material properties and reducing waste. This approach not only enhances the environmental performance of building components but also supports the creation of structures that are both functional and aesthetically pleasing, meeting the demands of modern sustainable architecture. Among the projects, the use of three-dimensional 3D printing with natural fibre-reinforced polymers presents a streamlined process that integrates fiber impregnation, material deposition, and curing in a single step. This consolidated approach enhances the efficiency and effectiveness of fabricating complex structures, rendering it highly promising for large-scale applications, as showcased at the Biennale of Architecture 2023 (Fig. 2.2 right), for which, within the 3D Natur Drucken industrial research project, a mushroom column was developed by the assembly of quadrangular curved panels, that are bioprinted with differential thickness to reach the optimum load-bearing capacity (Fig. 2.3).

Tailored Fiber Placement (TFP) has also been developed by the same research group led by Hanaa Dahy, as it is an advanced digital embroidery technique that creates intricate textile preforms by accurately positioning natural fibers in alignment with the intended force flow. Originating from the textile industry, TFP technology has been adapted for use in building construction, where it enables the fabrication of highly efficient, structurally optimized components. This method ensures precise

Fig. 2.3 3D Natur Drucken project developed by BioMat and exhibited at the Biennale of Architecture 2023 in Venice

fiber orientation and placement, resulting in materials that are both lightweight and robust, and suitable for demanding structural applications. This technique has also been developed for robotic assembly and showcased with façade panel production [9].

For prefabricated construction, AM can also be adopted for on-site manufacturing facilities and specialized labs, and it can facilitate the 3D printing of building envelopes and structures in separate parts that can then be assembled on-site. Common equipment consists of robotic arms, gantry-crane-based 3D printers, and medium or small prefabrication machines. In contrast to large-scale printers, printers commonly used for developing concrete prefabrication components can reach speeds of up to 0.150 m/s and thinner layer thicknesses of up to 60 mm. This makes it possible to create complex shapes more accurately and with less material. However, the drawback of longer printing times arises when the geometry becomes more complicated and, therefore, requires a reduction in printing speed.

2.1.3 3D Printing for Building Production

Realizing full buildings with AM technologies is today still a challenge, given that the technology is still not mainstreams, and it is not democratized. However, some experimentations have been taking place, with promising technological results. Among them the TOVA house developed by the Institute for Advanced Architecture of Catalonia (IAAC), is a 3D printed earthen small house, developed and fabricated to demonstrate the possibility to democratize 3D printing through the adoption of local material. What makes this application of special interest is not only the formal architectural result, which sees this as a one room house with earthen perforated wavy walls completed with a bi-directional wooden roof of exposed rectangular beams, but the technological approach. In their work, they have linked digital design with simulations, where these last have been tuned on the results of physical prototyping, testing and sensors, to achieve a good prediction and control of the 3D process and its final result. This approach has been applied to the TOVA house, which has been realized in a forest near Barcellona with locally sourced materials and has allowed to develop 3D printed wavy walls that variate their geometries to accommodate water protection, thermal performance, ventilation, as well as to integrate carpentry details, furniture and to support the wooden roof [10]. Water resistance is of particular importance in earthen buildings, as they can be vulnerable to water absorption. Therefore, to overcome this criticality, they 3D printed a formwork with the same curvilinear geometry as the rest of the walls and filled it with a mix of earth and geopolymers, the latter coming from recycling metal waste, in order to provide an appropriate water resistance to the base of the building. In the rest of the walls, specific offsetting of layers was implemented to reduce water absorption. Following the logic of adopting locally sourced material, the soil excavated for the foundation was used for 3D printing the house walls, and sisal fibres were added to the mix to achieve the required mechanical strength. Being realized in a Mediterranean country, appropriate thermal performance was achieved by variating wall thicknesses and properly tuning interior voids. For natural ventilation they used a combination of voids, and an embedded wooden case that can open and close as required. The result is a nine square meter building with the performance of a full house and the quality of details proper of a successful vernacular architecture.

2.2 Robotics in Construction

In an industry heavily reliant on manual labour, the adoption of automation is seen as a promising solution to various challenges. Automation has the potential to enhance productivity and quality, while simultaneously reducing the dependence on manual labour and minimizing the occurrence of accidents [11]. The current landscape of robotics in construction, as described by Bock [12], is characterized by the emergence of new technologies stemming from traditional construction practices. These technologies initially face performance limitations due to technical, organizational, and economic barriers, as well as limited integration within an industry still dominated by established technologies. However, the renowned interest in prefabrication, reframed in the MMC umbrella, has the potential to facilitate the development, diffusion, and adoption of automation in construction.

In the domain of construction robotics, it is crucial to design systems that can be seamlessly integrated with existing construction processes. This involves aligning the capabilities of robotics with the specific requirements of construction activities.

Despite ongoing research efforts to develop robots that closely emulate human capabilities, the physical attributes of robotics technology remain open-ended and subject to unpredictable advancements [13].

Given the growing stage of automated construction, there is a need for a flexible framework that accommodates evolving technological innovations without imposing rigid constraints. It would be important to develop a solution-neutral framework that can be rooted in fundamental robotic principles, which can be used for specifying, planning, and executing building element assembly to achieve desired construction systems [14, 15]. While these foundational concepts have long been studied in the field of automation, creating a formalized framework that transcends specific solutions could facilitate interdisciplinary collaborations and guide in navigating the complex and dynamic landscape of automated construction.

In an attempt to facilitate navigation of the landscape of automated constructions, and to help in identifying what are the main streams of research that can and could in the future enable further adoption of automation at the scale of building components and building systems, a four-stream framework is proposed, which look at: *a.* robotic platform, which considers the overall physical attributes of the robot; *b.* construction planning and end effectors, which involve the strategic planning of various automated construction procedures, along with the design and development of connection systems between parts and the corresponding robotic end effectors, ensuring meaningful interaction between the robot and the components; *c.* process monitoring, which includes the use of software to leverage sensor data for error correction and provide human operators with insights into the system's current, planned, and past actions; *d.* human–robot collaboration. In the following sections, robotic platforms, end effectors and process monitoring will be discussed more, to highlight their importance for the automation of construction and to discuss the potentialities and applications in the current building construction and research trends.

2.2.1 Robotic Platforms

Robotic platforms for construction represent a cutting-edge technological innovation that holds great potential to revolutionize the construction industry. These platforms encompass a wide range of robotic systems specifically designed to perform tasks traditionally carried out by human workers in construction settings. These platforms can be utilized for various construction activities, such as bricklaying, 3D printing of structures, material handling (Fig. 2.4), site monitoring, and even autonomous vehicles for the transportation of materials. They can be equipped with advanced sensors, actuators, and intelligent control systems that enable them to operate autonomously or semi-autonomously.

One of the key advantages of robotic platforms in construction is their ability to improve efficiency, accuracy, and safety on construction sites. By automating certain tasks, these robots can increase productivity, reduce errors, and minimize the risk of accidents, thereby enhancing overall project outcomes. Moreover, robotic platforms

Fig. 2.4 Robotic platform for the assembly of wooden facade system

have the potential to address labor shortages and rising labor costs in the construction industry by complementing human workers and taking on repetitive or hazardous tasks. They can also work around the clock, thereby potentially accelerating project timelines.

As the field of robotic platforms for construction continues to evolve, ongoing research and development efforts are focused on enhancing robotic capabilities, integrating these systems seamlessly with existing construction processes, and expanding the scope of applications within the industry. These advancements are expected to drive innovation, efficiency, and sustainability in construction practices for years to come.

Robotic platform for building construction can primarily employ either up to seven-axis robots or Cartesian robots. The most commonly utilized type of robots are arm-based robots, which emulate the movement of a human arm across various degrees of freedom. A degree of freedom refers to an axis along which the robot can move, with each rotational joint contributing one degree of freedom. For a rigid body to achieve complete freedom of movement within a three-dimensional space, a minimum of six degrees of freedom is required. Arm-based robots typically possess more degrees of freedom than other robotic platforms, enabling them not only to position an object in 3D space but also to orient it, a critical capability for assembly tasks. This versatility allows a single robot to perform a wide range of functions (Fig. 2.5, and [16]). However, a limitation of these robots is their often restricted workspace, which is typically confined to a hemispherical area determined

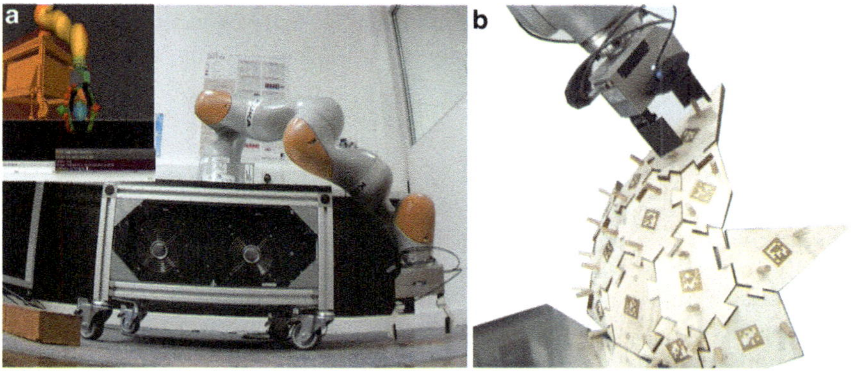

Fig. 2.5 Robotic assembly of curvilinear surfaces: **a** Pick and place of a curved masonry wall, with reachability study of a robotic platform made of one 7 axes Kuka Robot; **b** pick and place for the assembly of a self-standing wooden curved surfaces, as developed for the "Set-in Transition" design project commissioned for the Collateral event at the Biennale of Architecture 2023

by the robot's maximum extension [17]. This constraint can be partially mitigated by incorporating locomotion capabilities, thereby expanding the operational area.

Cartesian robots are frequently employed in assembly processes that involve layered construction. These robots are characterized by their limited capability to control the orientation of the end effector. The advantages of Cartesian robots include their relative ease of programming, mechanical simplicity, and the ability to support larger payloads due to their full-frame chassis, as opposed to a single arm bearing the entire weight. Unlike Cartesian robots typically used in other industries, such as 3D printers, using a full-frame chassis to automate construction production, can become cost-prohibitive, can occupy substantial space during transit, and require significant time to assemble, thereby negating many of its inherent advantages and preventing fully automated operation. Nevertheless, Cartesian robots in construction retain the benefits of easier programming and mechanical simplicity. Similar robots are used in bricklaying, mimicking the way a human would take and stack bricks.

There are also hybrid approaches, such as the one employed by Selective Compliance Articulated Robot Arm (SCARA), that can also allow rotation, and hence the deposition and installation of curved elements, as used, for instance, in the 3D-printing Clay project [18].

A good example of a robotic platform developed for prefabricated housing systems is provided by Reframe Systems, which utilizes design automation, modular building components, and digital manufacturing within robotic microfactories. The company set up by Engineers previously working at Amazon Robotics, has employed the principles of flexible automation, using readily deployable robotic work cells, primarily robotic arms, that are low-cost, reconfigurable, and that can be distributed to assemble prefabricated walls and floors, up to full volumetric housing units. The currently developed robotic platform can frame and connect boards, and with the development of movable platforms, they can also perform repetitive actions such as drywall, finish taping, and finishing applications. A more comprehensive overview of this prefab system approach is provided in the exemplary cases chapter of this book.

2.2.2 End Effectors

The end effector is a critical component in the automation of construction and assembly processes, serving as the primary tool that interfaces the robotic platform and the construction system. Its design must be intricately aligned with the specific parts and joints of the structure to ensure effective performance. The development of end effectors can follow various methodologies, each tailored to the unique requirements of the task at hand. End effectors can be either designed and fabricated for specific applications, or generic ones can be adapted for tailored uses. The customization of end effectors to suit specific applications is a well-established practice in robotic systems across various industries. For instance, some end effectors may be specifically engineered to perform particular tasks with high precision, reflecting the diverse and specialized nature of robotic applications. The design considerations for end-effectors often include factors such as the type of material to be handled, the required precision and dexterity, and the environmental conditions under which the robot will operate.

The end effector, indeed, must be designed to work well with all tasks it needs to complete. Often in construction, this will mean a mix of gripping items and fasteners as well as using tools on the fasteners to tighten them. One approach is to make a specific but versatile end effector that can be very adaptive in the set of situations that could present during the assembly of a specific construction item or system. Such an end effector, for instance, could be designed to grip and manipulate panels easily by resembling human hands. Alternatively, it is possible to adopt more generic end effectors (Fig. 2.6), that can perform some generic tasks, and develop instead specific construction joints that can be designed to be handled by the generic hand effector, while performing the required structural function [19]. When multiple robots are integrated into a platform system, then tasks can be distributed among the multiple robots, and hence, each robot end effector can be designed to perform specific tasks. This can allow construction complexity to be distributed across a multi-robotic platform, with the benefit of simplifying robotic tasks, to the expenses of programming multi-collaborative robots. A good example is for instance provided by Willmann [20] who investigated the impact of implementing robotics into earlier design stages and developed a method to fabricate a reciprocal timber structure through robotic assembly, requiring minimum human interaction. They specifically tuned a 6-degree-of-freedom robot arm to grab, manipulate, orient, and position a series of timber components in a spatial configuration to assemble a reciprocal mushroom column. Through a series of end effectors, the robot grabs timber sections, cuts them to the required lengths, and at specific angles, according to a CAD design, and positions them in 3D space. The final connection of each piece was completed through nail guns. This experience provides an example of how integrating end effectors design

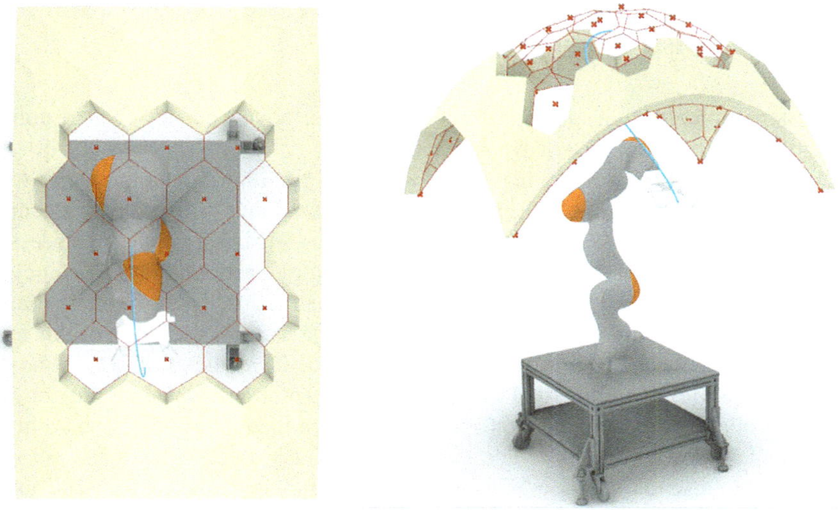

Fig. 2.6 Pick and place of voussoirs for the assembly of a shell system, as developed within the EPSRC research project "Exploration of parametric design and robotic assembly of shell structures"

and planning into the design process, can significantly enhance the efficiency and effectiveness of robotic platforms in automated construction.

2.2.3 Process Monitoring

Process monitoring has two significant aspects: human process monitoring and internal computing process monitoring. Among them, human process monitoring is the most relevant for this book's audience. Human process monitoring intends to provide the human with all the relevant data and actions a robot could and/or will perform. It allows for intervening in robotic planning and manipulation, as well as for optimizing robotic sequences, and can provide real-time information. This monitoring can be done by mobile end devices, which are easily readable by, also, less expert operators. The mobile device can be very efficient in displaying both partial and final 3D images of the completed construction, and, hence, providing info about reachability, space constraints, and task sequencing. Among the human process monitoring, also virtual and augmented reality can be considered, as they can provide a fully immersive experience. When the monitoring system is connected to a workstation, which collects sensory data, it could also be possible to understand and evaluate the difference between programmed and error-corrected robotic paths, and hence, intervene.

2.3 Augmented and Virtual Reality for Design Exploration and Construction Monitoring

In contemporary design and construction practice, advanced design and control technologies are finding extensive applications.

Virtual Reality (VR) is having a huge diffusion in the last decade, thanks to its peculiar ability to provide multisensory experiences within a virtual environment, as well as the possibility of integration with numerous BIM software. It is necessary to consider that in the last two decades, VR technology has found practical application in the most disparate fields, going beyond construction, from flight simulation applications to medicine, to cite just few [21]. The increasing diffusion of this technology in the field of Architectural Engineering and Construction (AEC), finds its main application in architectural design to: (1) interpret and visualize projects, (2) develop collaborative design, and (3) develop design maquettes [22]. The ability to modify and interact with space and three-dimensional objects, makes VR not only suitable for presentation and verification but also for the actual design and modification phase of the project. When compared to real scenarios, immersive virtual environments (IVEs immersive virtual environments) are particularly effective in simulating spaces, materials, and light [23, 24]. Therefore, they are tools specifically adapted to dialogue with all the project actors and facilitate the collection of feedback and evaluation of the degree of satisfaction of any final user, including non-technical individuals.

It is possible to divide the VR tools into three main categories: (i) fixed displays, which include rooms (quarries) in which the user is surrounded by three-dimensional screens, (ii) headsets, a category that also includes cardboard supports (google cardboards) in which the smartphone generates the stereoscopic image; and (iii) portable screens, such as smartphones and tablets [22]. There are numerous applications of VR technology in different contexts that highlight its effectiveness. They can be used, indeed, for design co-production, and/or for construction.

Looking at VR applications for design co-production, for instance, Aldabra et al. [25], experimented and compared the use of VR technologies with other architectural project visualization techniques, during the design phase of two refugee camps. For the scope, they developed VR projects of future potential camps and evaluated them with refugees. The results demonstrated the effectiveness of the tool within co-production contexts as well as the ease of use by a large number of participants. The accessibility of the VR application and the capacity to attain a larger number of feedback, also from non-technical participants, has also been demonstrated by Iuorio and Russo [26], who developed an immersive virtual reality experience to interrogate and verify the spatial arrangements of future post-disaster neighborhoods and buildings in Ecuador (Fig. 2.7). The work tested the capacity of immersive VR to enable experiencing both urban environment and domestic spaces, to evaluate the level of acceptance of urban and building design proposals by hazard survivors, who voluntarily participated in the in-person experiment. The work aimed to challenge

Fig. 2.7 VR development for interrogating the social acceptance of urban and architectural proposals within the research project NOVA VIDA

the current typology of post-disaster settlements, and the current typology of mass-produced post-disaster buildings, to pose at the center the long-term well-being of the inhabitants. The VR scenarios were tested in a series of workshops with representatives from current post-disaster settlements, NGOs, civil servants, and risk experts, and demonstrated the facility of communication with a wide range of stakeholders and the efficacy of experiencing intangible spaces and building technologies. Further studies have also demonstrated that the accessibility of VR can even grow with the integration of augmented reality (AR) [27].

VR and AR are also growing in popularity with adoption in construction processes, as they enable the simulation of tasks such as production, transportation, handling, and assembling of building components. Allowing real-time simulations, they can facilitate predicting assembly sequences and evaluating what components assemble first and in what position. The AR environment can allow team members to evaluate simulated assembly sequences and analyze different construction methods. AR has, for instance, recently been used for the layering of bricks for a think vault at the Biennale of Architecture of Venice 2024, when a team of academics, researchers and bricklayers tested the efficacy of AR to facilitate the layering of an intricate pattern of three layers fast setting bricks [28].

VR has also been used, specifically, for the assembly of a building starting from prefabricated components [29]. In that experience, the design space has used a virtual environment, which included a library of 3D prefab components, that the clients could use to design the building, while at the same time, could check the feasibility of the construction/assembly through automatic constraint recognition. The project also used an integrated virtual reality display to interoperate with the simulated construction assembly, which can be used by the project manager to plan and optimize the entire construction process, and the corresponding material, components, and budget resources.

2.4 Conclusion

The advancement in AM is providing the AEC sector with the opportunity to overcome technical and process limitations of traditional construction methods to tailor material, and building components, geometries and performances to address customization of aesthetic, mechanical, thermal, and acoustic requirements while enhancing resource efficiency through optimization processes. This chapter reflects on material advancements in AM material based on clay, steel (WAAM processes), earth and concrete (3DCP), and presents their applications at both the component and full building scale. The AM prototypes of building components cover bricks, columns and façade systems demonstrating a plethora of opportunities for a more creative built environment, in which the objective of lowering carbon emissions can be achieved by material, shape, and topological optimization, as for the case of S-bricks that can be assembled in striking facades system and/or breathable facades, or the Nature Drucken mushroom columns that unlock opportunities for bio-printed tiles with variable thickness, to withstand mechanical requirements, while creating a natural forest effect.

Robotics on the other hand is advancing the capability of the AEC sector by taking on repetitive or hazardous tasks, while also providing the opportunity to advanced manufacturing to be employed at a large scale. However, the field of automation in construction is so broad, that to navigate it and to enable further adoption at the building component and full building scale, this chapter proposes a four-stream framework, which encompasses robotic platform, planning, effectors, monitoring, and collaboration, as they are considered as the main streams requiring continuous research, in order to be implemented efficiently in the interdisciplinary landscape of AEC. Among the identified streams, it poses attention, in particular to robotic platforms and effectors, as they can enhance robotic capabilities and allow integration with construction industry. In an ideal world where robotic manufacturing and assembly are considered from the inception of building design, the domains of robotic, architecture and building construction can influence each other opening up to formal and technical languages still to be seen.

References

1. ISO/ASTM 52900:2021 Additive manufacturing—General principles—Fundamentals and vocabulary
2. ASTM F2792 (2012) Standard terminology for additive manufacturing technologies. ASTM International, West Conshohocken, PA
3. Ralph B (1925) Method of making decorative articles. United States, US1533300A
4. Tomar B, Shiva ST (2022) Nath review on wire arc additive manufacturing: processing parameters, defects, quality improvement and recent advances. Mater Today Commun 31
5. Lu B, Weng Y, Li M, Qian Y, Leong KF, Tan MJ, Qian SA (2019) Systematical review of 3d printable cementitious materials. Constr Build Mater 207:477–490

6. Bianchi I, Volpe S, Fiorito F, Forcellese A, Sangiorgio V (2024) Life cycle assessment of building envelopes manufactured through different 3D printing technologies. J Clean Prod 440
7. Cruz PJS, Camões A, Figueiredo B, Ribeiro MJ, Renault J (2020) Additive manufacturing effect on the mechanical behaviour of architectural stoneware bricks. Constr Build Mater 238
8. Laghi V, Palermo M, Gasparini G, Trombetti T (2020) Computational design and manufacturing of a half-scaled 3D-printed stainless steel diagrid column. Addit Manuf 36:101505
9. Dahy H (2023) Materials as design tool: digital fabrication strategies for sustainable architecture, technology. Arch + Des 7(2):153–158
10. Dubor A, Cabay E, Tayoun Y, Carrasco O, Carnevale E, Chamorro E, Huygues V (2024) 3D-printed earth architecture design approach for a performative habitat. Fabricate 2024
11. Neelamkavil J (2009) Automation in the Prefab and modular construction industry. In: Proceedings of the 26th annual international symposium on automation and robotics in construction 2009. University of Texas Press, Austin, pp 299–306
12. Bock T (2015) The future of construction automation: Technological disruption and the upcoming ubiquity of robotics. Autom Constr 59:112–121
13. Ruggiero A, Salvo S, St. Laurent C (2016) Robotics in construction. Worcester Polytechnic Institute, Worcester
14. Everett JG, Slocum AH (1994) Automation and robotics opportunities: construction versus manufacturing. J Constr Eng Manage 120(2)
15. Liu Y, Choi J, Napp N (2019) Planning for robotic dry stacking with irregular stones. Department of Computer Science and Engineering, University at Buffalo, NY, Buffalo
16. Iuorio O, Wilcock S, Korkis E (2024) Design for deconstruction through digital fabrication of thin spatial systems. In: International conference coordinating engineering for sustainability and resilience. Springer Nature Switzerland
17. Wilcock S, Boyle JH, Dogar MR, Iuorio O (2022) Automated robotics agents for assembly-aware design of shells. In: Proceedings of the international conference in structures and architecture, ICSA 2022
18. San Fratello V, Rael R (2020) Mud frontiers. In: Menges A, Sheil B, Glynn R, Skavara M (eds) Fabricate. UCL Press, London, pp 22–27
19. Korkis E, Dogar M., Iuorio O (2023) The role of parametric design in the robotic assembly of dry-constructed shell structures. In: Proceeding of the IWSS 2023
20. Willmann J, Knauss M, Bonwetsch T, Apolinarska AA, Gramazio F, Kohler M (2016) Robotic timber construction—expanding additive fabrication to new dimensions. Autom Constr 61:16–23
21. Kim MJ, Wang X, Peter ED, Love, Li H, Kang S-C (2013) Virtual reality for the built environment: a critical review of recent advances. J Inf Technol Constr (ITcon) [online] 18(14):279–305. http://www.itcon.org/papers/2013_14.content.06510.pdf
22. Zhang Y, Liub H, Kanga S-C, Al-Husseina M (2020) Virtual reality applications for the built environment: research trends and opportunities. Autom Constr 118
23. Heydarian A, Carneiro JP, Gerber D, Becerik-Gerber B, Hayes T, Wood W (2015) Immersive virtual environments versus physical built environments: a benchmarking study for building design and user-built environment explorations. Autom Constr 54:116–126. https://doi.org/10.1016/j.autcon.2015.03.020
24. Moscoso C, Nazari M, Matusiak BS (2022) Stereoscopic images and virtual reality techniques in daylighting research: a method-comparison study. Build Environ 214:108962. https://doi.org/10.1016/j.buildenv.2022.108962
25. Albadra D, Elamin Z, Adeyeye K, Polychronaki E, Coley DA, Holley J, Copping A (2020) Participatory design in refugee camps: comparison of different methods and visualization tools. Build Res Inf:1–17. https://doi.org/10.1080/09613218.2020.1740578
26. Iuorio O, Russo M (2022) Future scenarios for housing (re)settlements in Ecuador. Arch Struct Constr 711–722(2):4
27. Wang X, Kim MJ, Love PED, Kang S-C (2013) Augmented reality in built environment: classification and implications for future research. Autom Constr 32:1–13. https://doi.org/10.1016/j.autcon.2012.11.021

28. Gaspar O, Paris V, Herrea F, Oval R, Olivieri C, Al Asali W, Beghini A, Adiaenssens S (2023) The new role of craft: Angelus Novus vault. In: Proceedings of the 43rd annual conference of the association for computer aided design in architecture, ACADIA 2023, vol 1, pp 232–241
29. Murray N, Fernando T, Aouad G (2003) A virtual environment for the design and simulated construction of prefabricated buildings. Virtual Reality 6(2):44–256

Chapter 3
Engineering the Systems: The Role of Research and Development

Abstract The prefab market is extremely competitive, and recently its strength is based on the capacity to offer high performances (structural and environmental) while containing the execution times and the costs. The sector's robustness is increasingly contingent upon the ability to innovate and integrate advanced systems that enhance overall efficiency and sustainability. Entities that excel in this market are typically those that allocate substantial resources towards research and development (R&D), enabling them to develop and implement groundbreaking technologies and delivery approaches. This chapter presents the pivotal role played by R&D in three recent emblematic MMC construction systems. They represent research and development in: *a.* volumetric light gauge steel system for large-scale manufacturing, *b.* wooden-based cassette system to be delivered through flying factories, *c.* a platform approach to be applied for a variety of functions and contexts. For each of them the idea, the main research strands and exemplary applications are presented.

Keywords Modular systems · Flying factories · Platform approach · Construction industry

3.1 The Case for Research and Development

Modern Methods of Construction (MMC) are another way the public sector is referring to prefabrication. MMC is spreading in the Global North due to the necessity to build, in a short time, a large volume of buildings to house and provide the necessary services to a growing global population, while at the same time achieving high standards in terms of structural safety and energy performance [1, 2]. The historical cycles of rising and falling of prefabricated systems, which were initially referred to as industrialized buildings [3], then system buildings [4], and more recently as offsite prefabrication [5], has hopefully made it clear, that to allow for the construction sector to improve the level of industrialization to potentially speed up processes and help with productivity, the new prefabricated systems must demonstrate their performance and their capacity to deliver building systems at required standards of

O. Iuorio, *From Mass Prefab to Mass Customization*,
PoliMI SpringerBriefs, https://doi.org/10.1007/978-3-031-73327-7_3

structural safety, fire safety, and energy performances, while also making places that can nurture the community. Current codes properly regulate traditional construction systems, while they quite often delegate to "design assisted by experimental testing" for non-conventional construction systems. This is certainly the case for the definition of the structural and fire performance of any non-traditional system, which through adequate testing needs to demonstrate the safety requirements. To understand the importance of research in the growth of innovative prefabricated systems, the following sections present the product development of three systems, which have three distinctive underpinning objectives. The first one is demonstration of a company aiming to produce a volumetric system with high speed, large investment capital and for a specific need (low-rise housing). The second one, instead developed by an architecture studio spin-off, aims to democratize to some extent the use of prefabrication, through cassette systems that have a digitalized design process but could be manufactured anywhere through flying factories, and in this way allow appropriation by users and or manufacturers. The third system, instead propose a platform approach that could be potentially used for any application, and that aims to foster a network of industries that could interoperate.

3.2 Enabling Mass Manufacturing of Volumetric Lightweight Steel Houses Through Product Optimization

Lightweight steel systems, also called Cold-formed steel (CFS) structures are gaining in popularity as off-site systems employed in particular for low and mid-rise housing, as well as partition and infill walls in more traditional construction systems [6].

The system comprises CFS members that are assembled through mechanical connections to form horizontal diaphragms and lateral resisting walls. To achieve the diaphragm action and the lateral capacity, the steel members need either to collaborate with steel bracing systems (all-steel design), or they can collaborate with sheathing panels (sheathing braced design approach) [7]. Among them, the sheathing braced design is of particular interest since it allows to obtain a required lateral strength capacity with a reduced amount of steel, and hence, allows for obtaining lower environmental impacts. However, while the sheathing braced methodology is codified in USA (AISI S400), it still lacks of appropriate codification in Europe, and, therefore design assisted by testing is required. Specifically, a system was developed as part of a Knowledge Transfer Partnership (KTP) between the University of Leeds and Ilke Homes and led by the Author of this book. The research project titled "Optimization of cold-formed steel systems for large scale manufacturing of modular houses" was funded by Innovate UK and aimed to develop an optimized cold-formed structural system utilizing novel design and manufacturing techniques to significantly improve the structural performance, safety, and sustainability of modular homes. To achieve this intent, the work aimed to shift the structural design of this modular CFS home from X bracing design to sheathing braced, with the intent to reduce the amount of material, facilitate the manufacturing process, and reduce the embodied carbon. The study adopted a methodology that integrated manufacturing and assembly strategies from the product development stage. In this approach, materials and components were chosen, prototyped, and tested, over 3 years (2019–22) to optimize both mechanical and environmental performance [9].

The experimental work is of particular interest for this book, as it testifies to the necessity of research and development for the spreading of more technologically advanced systems in construction.

To optimize the structural system, reduce the amount of material used, and consequently reduce the associated embodied carbon while respecting the lateral performance of the building, and facilitating the manufacturing and assembling of the building, a large experimental campaign including micro, meso, and macro tests was planned and performed. Specifically, micro tests included 32 tensile strength tests for the steel material and 20 lap-shear tests on screws, while mesoscale tests were developed to define the shear strength of connections between steel profiles and either oriented strand boards or cement board panels. Wall tests were developed in two phases, with a first phase to characterize the shear behaviour of 2.400 × 2.974 m fully sheathed walls, and a second phase to understand the behaviour of walls with openings. Figure 3.1 reports the overall experimental campaign.

In the first phase of the wall tests, four full-scale specimens without opening were tested under horizontal loads to characterize experimentally the lateral strength and stiffness of typical ground floor (2400 mm × 2974 mm) and first floor (2400 × 2926 mm) houses. Therefore, every wall was made of CFS studs, tracks, and

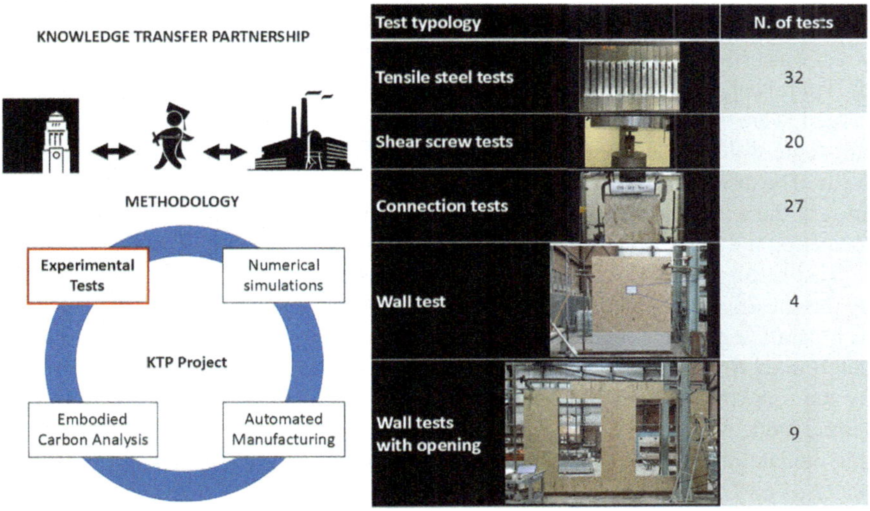

Fig. 3.1 Overall methodology and experimental campaign applied by the knowledge transfer partnership between the University of Leeds and Ilke Homes

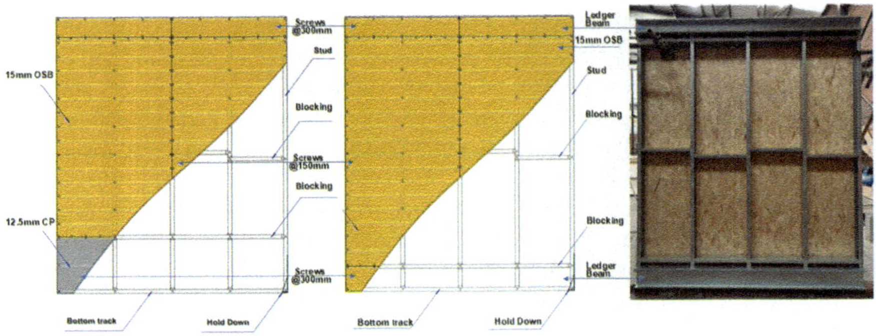

Fig. 3.2 Geometry of the walls tested in the first stage

blocking profile as schematized in Fig. 3.2, made from steel grade S450 and having a thickness of 1.6 mm.

The peculiar wall configuration was due to the necessity of using three different sheathing panels across the height of the walls, to have cement-based panels (CP) in the lower part of the walls to guarantee the required structural and humidity performance, and two oriented strand board (OSB) panels at the central height and at the top of the walls (also in this case to guarantee the required structural performance, while also facilitating the manufacturing process). In all walls, structural sheathing panels were only placed on one side of the walls, to leave the freedom to adopt a variety of sheathing systems for the inner side of the walls to reach the requested thermal and fire performances.

The evaluation of the screw typology and spacing to be adopted as a connection system between the CFS profiles and the sheathing panels across the load-bearing walls was critical, as the number and typology of screws are strictly responsible for the lateral capacity of the load-bearing wall system. The mesoscale connection tests, indeed, were carried out to quantify the shear capacity of the selected screws, and the full-scale wall tests were performed to evaluate the lateral capacity of the designed walls. The tested walls also included Simpson Strong-Tie HTT5 hold-down are both ends of the walls, to avoid overturning during the testing, and perimeter beams were located in the inner side of the walls for, respectively, floor and ceiling cassettes.

Test results revealed that for all GF specimens, the wall collapse was governed by the sheathing-to-frame connections. At the global level, the steel frame deformed as a parallelogram with a consequent rigid rotation of the sheathing panels, which determined first the titling and pull-through of all the screws, which was followed by the cracking of the CP, and the edge breaking of the CP panels corners. This failure mode was in agreement with the predicted behaviour, and it was demonstrated that the lateral strength was higher than as required by the codes. Details of the first phase of wall experiments are published in Iuorio et al. [10]. These test results were then extended to other configurations through numerical analysis, that specifically allowed the prediction of walls' lateral capacity at varying sheathing-to-frame screws distance and sheathing heights. Based on the results of the numerical prediction,

Fig. 3.3 Factory production line: **a** high-speed panel system; **b** wall panel assembly line; **c** complete volume system

screw spacing and sheathing height were optimized to allow screws installation at high speed through an automated process (Fig. 3.3).

In the second phase of the full-scale tests, nine tests of 3 typologies of walls with openings were performed. Two walls were replicating typical ground floors (GF, 4800 × 2964 mm), with the GF F having two openings, corresponding to a door and a balcony for a total opening of 2100 mm, and the GF R having only one large opening of 2680 mm (Figs. 3.4 and 3.5). The third wall typology FF was representative of a typical first floor with a length of 4800 mm and a height of 2926 mm and had two openings for a total width of 2210 mm (Fig. 3.4). The walls were selected to represent the worst-case scenarios in terms of opening ratio (α) among those manufactured by the housing provider. It is important to notice, that, with the intent to reduce at the maximum the material waste generated during the manufacturing stage, OSB sheathing panels were continuous across the opening corners. This corner detail is certainly not commonly used in previous CFS applications.

As for the previous wall tests, in this second experimental phase, the CFS frame tended to deform in a parallelogram and the sheathing boards were prone to rotate. However, given the presence of ledger beams at the top and the bottom of the walls and the specific sheathing arrangements, which included full-height sheathing boards in the centre of the wall and short-height panels on the top and bottom part, the central part of walls were to experience larger deformation, with specifically the central sheathing panels being subjected to larger rotation. This determined pull-through of the screws predominantly around the edge of the central panels. In terms of strength and stiffness, when comparing the results between the two experimental phases, the walls with opening exhibited a lower strength and stiffness, which was primarily due to the presence of the opening. The results, indeed, showed that the opening mostly influence the stiffness of the walls. When comparing the FF with opening,

Fig. 3.4 Typologies of walls with opening tested during the KTP project

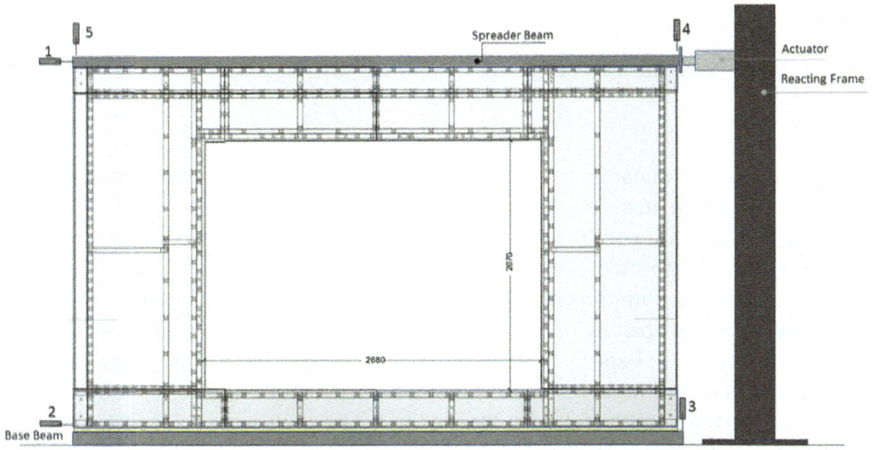

Fig. 3.5 Scheme of the Experimental test set-up with the indication of the reacting frame, the actuator for the load application, the spreader beam to distribute the loading across the wall, and the displacement sensors (from n.1 to 5) located for measuring the horizontal and vertical displacements

with a similar configuration but without opening, the stiffness decrease of about 6.4%. Reducing the screw spacing has a larger contribution on the shear strength of the walls. Indeed, as a result, the GF with the larger opening (GF R) has higher strength than the similar (GF F) but lower stiffness.

Overall the experimental campaign allowed to verify the possibility of using OSB and cement-based (CP) panels as structural bracing systems, for the developed volumetric housing system. It also specifically allowed to quantify the lateral capacity of walls with and without openings, and based on that developing a design procedure for what was called "the refurbished" system, which was the structurally optimized system. The optimization allowed a reduction of used steel by 10%, with a significant consequent embodied carbon reduction, as well as speed up of the wall productions.

In conclusion, the research project undertook material and component research, computational analysis, testing and design work to optimize the overall structural steel design with both static and dynamic loadings (occurring during lifting and transportation of the modules). Through carrying out this detailed analysis and testing, the KTP enhanced the performance, cost and simplicity of assembly which has improved the viability of the Ilke Homes business across the UK and potentially could enable entry into overseas markets. Moreover, through the research project, the industry has developed its in-house structural engineering capability by upskilling most of the design team on the outcomes of the analysis and testing. The KTP has also created new design parameters enabling all current and future products to benefit from the results of the KTP. This was embedded in the structural design philosophy, in the new designs, in their standard operating procedures, and design and manufacturing program.

3.3 Development of Off-Site Constructions for Social Innovation

Developing an off-site system that could be democratized, by distributing the ownership of the production, has been the intent of the MassBespoken system. As the name itself implies, MassBespoken is a user-centered building system combining the benefits of mass production with bespoke design. The system aimed to develop full houses from flat-pack wooden panels, made of oriented strand boards (OSB), which could be laser cut to be assembled in the factory through finger joints and mechanical connections to make tridimensional wall and roof cassettes, that could be easily handled to assemble full houses. Key requirements of this development were to allow flexibility, in terms of building layout, inclination of walls and roofs, volumetric articulation of the overall building, and flexibility in the on-site cassettes assembly process, to be done without requirements of cranes.

The process is digitally enabled starting from parametric models to create adaptable designs. The architects developed the concept, together with all the structural and construction details so, that being in a BIM environment, they could be sent to outsourced fabricators.

The primary parametric modeling was integrated with the structural analysis model, which automatically verifies the designed structures against structural codes and building specifications, and iteratively resizes the structural members until all elements are verified. The model allows them to generate G codes for CNC fabrication together with additional manufacturing information (Fig. 3.6). The bespoken nature of the developed platform can allow the modular prefabricated system to be adapted to personal needs, as well as to be adjusted for a variety of sites and land sizes, shapes, and gradients.

The household could potentially input into the 3D modeling and visualization processes of their bespoke self-build home, while the construction could be fabricated

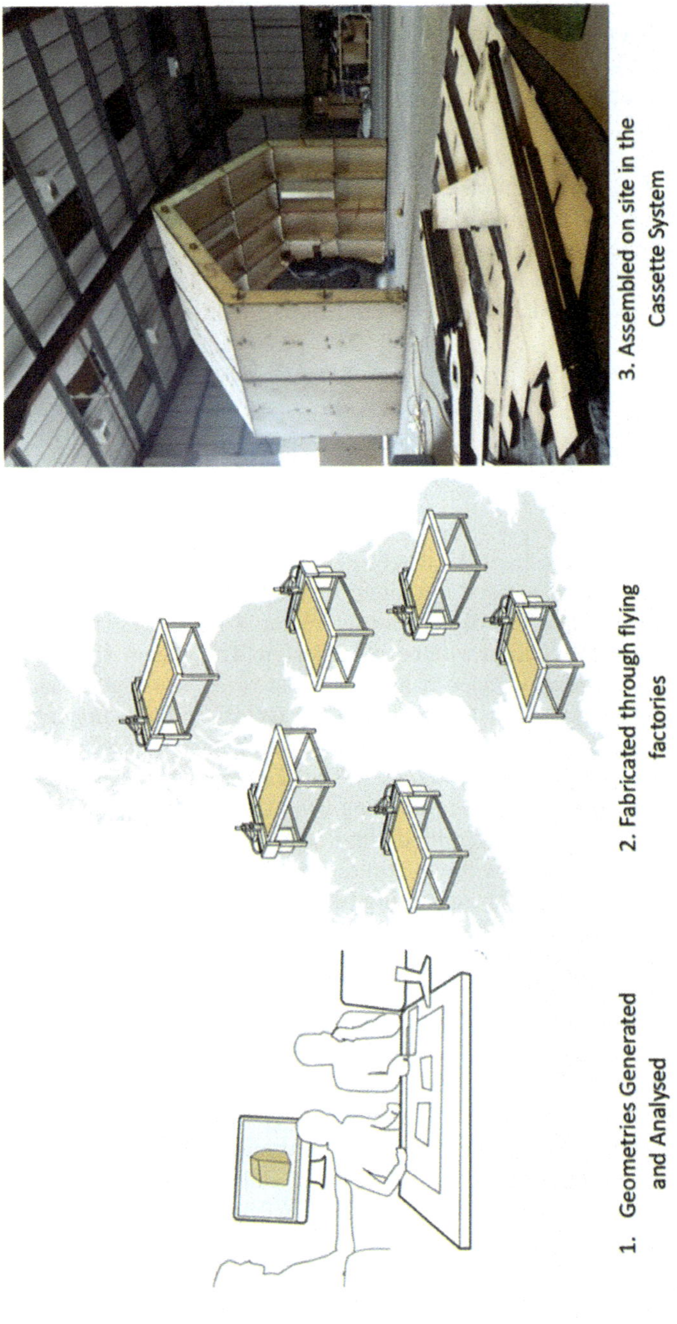

Fig. 3.6 MassBespoken design and fabrication concept (credited to MassBespoken)

through ‘flying factories’ (Fig. 3.7), which are temporary factory spaces that can be set up close to an assembly site and moved easily. They envisage, indeed, customers to design their homes on tablets, send fabrication details to fabricators, and bring CNC machines next to the construction sites so that most of the components could be fabricated close to the site, with the obvious effect of simplifying transportation burdens, reducing environmental costs and increasing localized ownership. This was done to oppose to the growth of large off-site facilities, which require huge factories, with very complex and generally large spaces, as well as expensive machinery to operate, and large forward investments. The system also claimed to require a less skilled workforce, which is an element not to be overlooked, in periods when skilled labours are rare and expensive. In an ideal case, the MassBespoken system could have been self-built, with cassettes lifted manually, allowing for the system to be affordable, without losing in performance and space quality.

MassBespoken was developed in collaboration with ARUP through multiple Innovate UK projects and a Royal Society Commission, which helped in defining the structural capacity of the system, developing the integration of the architectural and manufacturing digital workflow with the structural analysis, and allowed to develop the proof of concept, which was instrumental in getting the systems closer to market, within the MMC outskirt. As part of the experimental study, basic compression tests on OSB panels assembling were developed, and some of the connection systems were experimentally characterized. The research also identified structural constraints for the systems, mainly related to a range of opening sizes, and the allowed number of storeys. For the inter-story height, limitations coming from the OSB panels size were cleverly solved by staggering different height cassettes vertically and introducing the necessary moment connections to guarantee appropriate loading transfer paths.

One of the projects also focused on the role and possibility that distributive industry can offer, and what impacts collective making and ownership of tools can make.

Fig. 3.7 Flying factory credited to MassBespoken

Fig. 3.8 MassBespoken in Doncaster: construction stages and final building

Park Road was so far the main fully developed building for housing. The building is a two-floor detached house, which accommodates a flat on the first floor and a duplex over the first and second floors. The land belongs to the client, who was interested in trying new methods of construction (Fig. 3.8).

For the structure, a hybrid system was employed with a wooden frame and traditional wooden floor, with the envelope realized with MassBespoken cassette system. The requirement for a primary wooden frame for the structure was due to safety concerns, as the system was considered still as in part experimental at the time of the design. In other words, despite the quite large body of experimental work carried out in the research project, the acquired data were not considered sufficient to fully trust the load-bearing capacity of the system. The fabrication of the cassettes was developed by BlockBuild, which collaborated with MassBespoken to realize the systems at the required spec and developed the existing concept in a full real pipeline production. BlockBuild, itself, is a collective of fabricators that develop a series of digitally-driven wooden cassettes, for a large set of clients. BlockBuild fabricated the MassBespoken system for the Park Road development, relatively locally (100 km from the site), bringing with it social and environmental benefits.

The benefits and risks of this project were all shared by the three main actors (i.e. the architects, the builder, and the client) especially because this was a prototype. They jointly achieved a successful project. The collaboration between the involved stakeholders allowed to deliver the project on budget, at the expected quality, so that the units were lettable in a real market. This demonstrated the value of the approach bridging digital innovation with localised production, and affordability.

In conclusion, MassBespoken is an excellent example of a prefab system that is rooted in experimental research, and that brings an innovative way to think about prefabrication, by devising a cassette system, completed on-site, which aims to democratize prefabrication by allowing customers to input into the design, providing flexibility of solutions, allowing local fabrication of the components, opening the production to lower-skilled workforce and avoiding the necessity of large upfront investment for the production. The beauty of MassBespoken is that it is parametric and can be used on small and less lucrative sites. Their vision is for a network of small factories that could be resilient because they are networked and low-cost.

3.4 Development of a Platform Approach to Be Used in Multiple Contexts

The main intent of a platform approach is to be able to define a platform for which a series of standardized and interoperable components can be used, even if coming from a variety of suppliers. Therefore, the platform approach starts with the definition of a "kit of parts" that can be deployed for multiple building types.

This would allow interoperability of components across different sectors. Brydenwood initiated the definition of their platform by developing the idea of a "reference design", which they define as *"a beacon of efficiency and uniformity, offering design and construction professionals a new approach to achieving streamlined and flexible outcomes."* [11] The idea of reference design involves creating a centralized, highly optimized core design that can be customized, localized, and adjusted to fit unique sites and conditions. The triggering underpinning objective, for them, is to avoid developing a full new design from scratch every time, and instead, be able to use a reference design that can be tailored, and to some extent customized, for new necessities. A system of this type would allow reducing building volume, controlling variations between different constructions, allowing planning for manufacturing, optimizing material usage, and therefore reducing waste. From a productive and logistic perspective, a platform system could increase productivity, speed up installation, enable better control of the logistics, and potentially reduce transactional costs (Fig. 3.9).

Based on this objective, Brydenwood arrives to define a system that is standardized up to about 80% while keeping a 20% possibility to be adapted based on specific clients and site requirements. This would be to devise the attention and expertise of design and construction teams to focus on responding to the local, functional, and aesthetic requirements while using a reference design that is "*rooted in industry best practices*". For instance, a new hospital construction program in the UK is adopting the "reference building" design principles for the development of 40 new hospitals across the country.

Therefore, given the challenges of delivering quick, effective, and high-quality buildings, adopting reference design offers a solid foundation supporting efficiency while advancing the design and construction sector into a future of ongoing innovation and reduced procedures.

The concept of a platform and how could be devised in the construction industry was initially set in "Delivery Platforms for Government Assets: Creating a Marketplace for Manufactured Spaces" (2017), where they also identify 3 typologies, Platform I, II, and III based on the dimensional span of the structural system. Among them, Platform II dedicated to mid spans, between 6 and 9 m, appears to be most flexible to be applied for a larger number of building typologies (such as residential, commercial, school, and prisons, among others). The concept is based on a "carrier frame" that can be combined with a large range of complementary prefab products. This would allow many companies to develop envelope and internal prefab systems

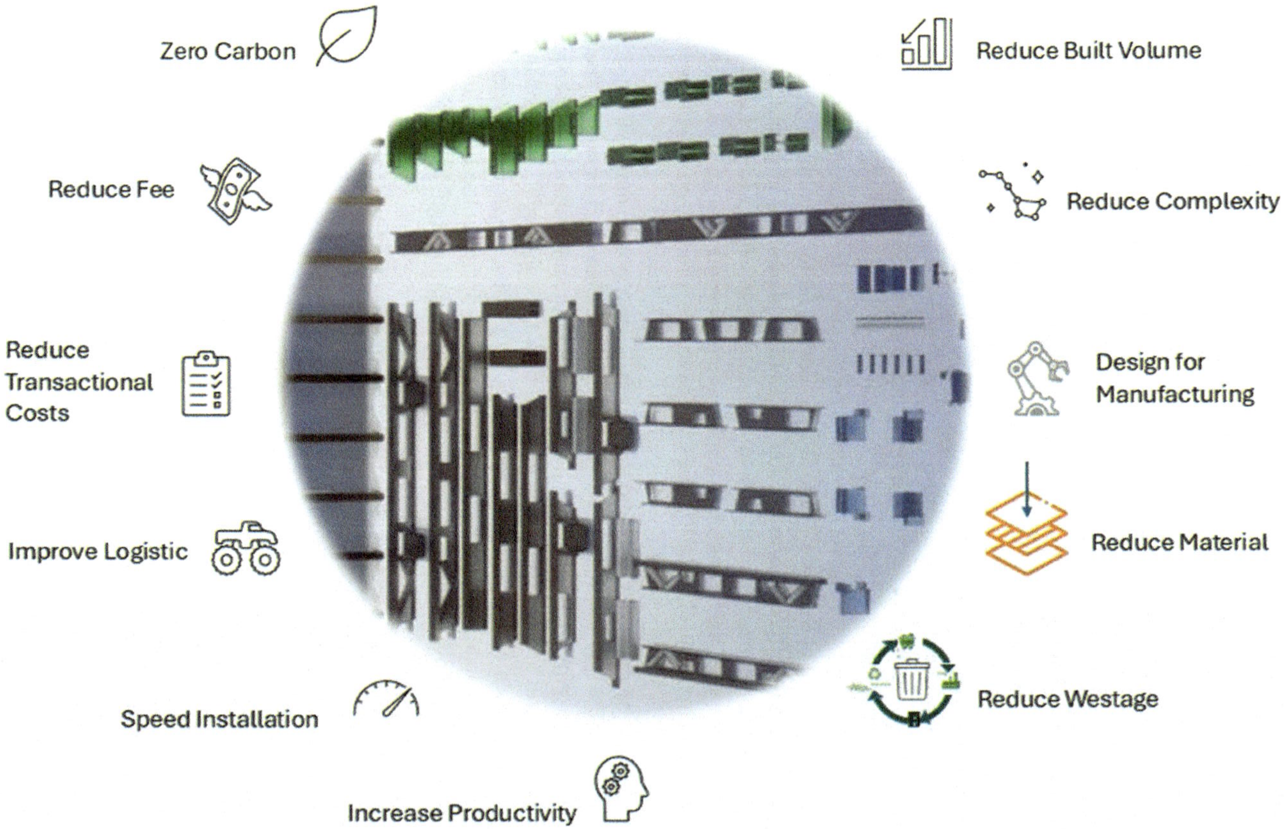

Fig. 3.9 Objectives of the platform system

that can meet independent energy and comfort requirements while being compatible with the structural carrier frame of the Platform.

The enabler of this vision would be a supply chain network of producers, from small to large companies, that find in a digital marketplace the opportunity to collaborate.

To arrive at the definition of the Platform systems, Brydenwood articulated their research in understanding the future needs for construction in the UK, defining the kit of parts, and testing them through scenario analysis and real case study constructions.

To provide evidence that Platform II, specifically could deliver a large portion of the future developments under planning by the UK government, they analyzed the data acquired by the Construction Innovation Hub, which looked at £50 billion of pipeline works across a series of UK Ministers and categorized those according to Uniclass categorization of individual spaces and found out that potentially Platform II could be used in 70% of the cases.

They investigated the role that superstructure, facade, mechanical, electrical, and plumbing services (MEP), and fit-out play in the overall construction and focused on the definition of the kit of parts for a low-carbon superstructure, aware that a lean and accurate superstructure, with already embedded possibilities for fixing points and interfaces with other building elements, can facilitate the adoption of other prefabricated systems, which potentially can already be mature in the current façade, and MEP markets. To define it, they analyzed typologies of applications, from housing to hospitals, administration, and education, and identified what are the most common and repeatable components that can be standardized, across the sectors. Based on that Platform I and II were defined and prototyped. In the same logic of analyzing performances of existing systems to then derive one most appropriate for multiple applications, Platform II has been developed as a hybrid system, which conjugates the accuracy, good compression performance, and flexibility to be connected to multiple subsystems of steel for columns, beams, and connections, with the good fire, acoustic and vibration performance of concrete, which is used for slabs. Specifically, they used standard steel components for columns and beams, and they customized the brackets so that they could only be installed in the correct position during the erection of the building, they also color-coded them to simplify assembly and make it obvious any potential error (Fig. 3.10). For the concrete slabs, they used composite thin slabs with preformed steel shutters and self-compacting concrete which is pumped across the floor and filling the shutters to obtain continuous diaphragms. In this way, they combine steel tensile capacities with the mass and fast construction of pumped concrete.

Therefore, the kit of part of Platform II is composed of permanent components (steel columns, beams, and brackets) and steel–concrete composite slabs; temporary reusable elements for the construction processes (shutters, temporary props, stillages), and processes and equipment for lifting.

The Forge is one the best completed examples of the platform approach, and for the quality of design has received multiple awards.

The Forge is the first commercial building developed with a platform approach and the design is very context-specific. The overall architectural design was developed

Fig. 3.10 Platform II kit of part (credited to BrydenWood)

by other architects, according to UGBC net zero carbon targets before Brydenwood entered the project. The original project was rationalized and turned into the "kit of parts" of Platform II, respecting the net zero carbon objectives. The design included automated configuration of digital components (Fig. 3.11). They used a combination of largely used construction software (as Revit and Tekla), in combination with algorithmic software routines with Grasshopper and Dynamo. The manufacturing processes included robotic cutting and welding of components. In this way, elements such as brackets were digitally designed and robotically cut, to reduce the possibility of human error and increase speed, productivity, and accuracy. The assembly process used repeatable on-site activities of lifting and fixing done through temporary reusable elements (such as formwork tables with pre-installed MEP fixing points, temporary props, and handrails) specifically developed for Platform II. The system for the Forge was first prototyped to test assembly sequences and many operations including the concrete pouring, as well as to test the capacity to reuse the designed temporary props efficiently (Fig. 3.12). The analysis of the design and assembling of the prototype allowed for refining the design and improving the on-site operations.

Once realized, The Forge demonstrated the capacity to integrate a platform approach with BIM technology, to realize a system precisely manufactured off-site and assembled with automated processes, while reaching the target Net zero requirements, and advancing the full overall supply chain towards construction 4.0.

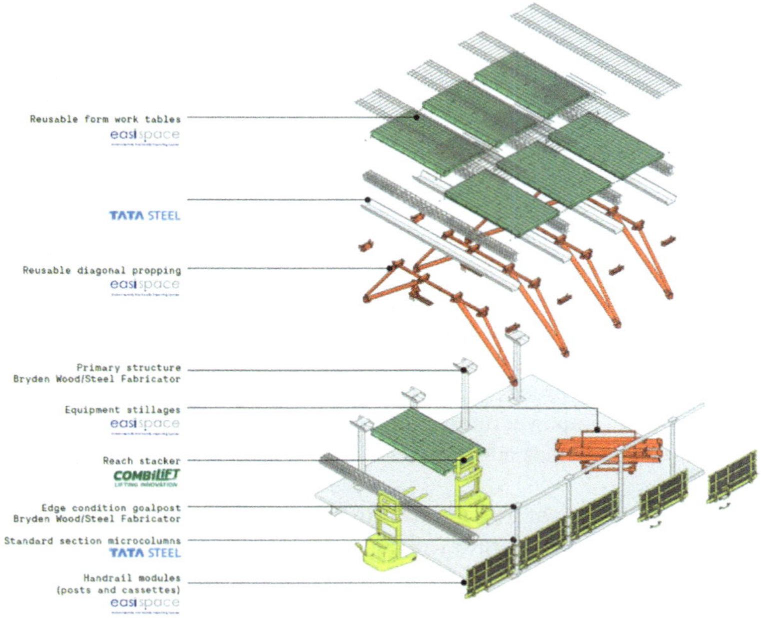

Fig. 3.11 Kit of part for the Forge (credited to BrydenWood)

Fig. 3.12 Prototyping in HMP Lindholme (credited to BrydenWood)

3.5 Conclusion

This chapter presents three exemplary cases of applied research for the development of three distinctive approaches to prefabrication. The first one emphasizes the importance and value of structural and technological experimental research for the development for manufacturing (DfM) of a volumetric system that is rooted in the steel industry but leverages academic knowledge to optimize the system and go beyond the codified practices. This produces the benefit of gaining better structural performance and technology quality, while reducing the amount of material, waste, and embodied carbon. The system is developed for mass manufacturing.

The second case, instead, uses research to develop a digital process for the bespoken design of a cassette system. The direct correlation between the digital process and the manufacturing process allows for the system to be distributed with flying factories, and hence, empower future users to be also the builders of their prefab buildings. This system combines customization and democratization of the construction process.

The third case uses a research approach to develop a kit of industrialized parts that can be interoperable across a variety of construction applications. The approach can allow a high degree of customization while using mass-manufactured components. A system of this type could enable the assembly and disassembly of a building multiple times, as well as mix-and-match the kit of components for diversified applications.

References

1. Iuorio O, Wallace A, Simpson K (2019) Prefabs in the North of England: technological, environmental and social innovations. Sustainability 11(14)
2. Iuorio O, Napolano L, Fiorino L, Landolfo R (2019) The environmental impacts of an innovative modular lightweight steel system: the Elissa case. J Clean Prod 238
3. Diamant RME (1964) Industrialised building: 50 International methods. Iliffe
4. Finnimore B (1989) Houses from the factory: system building and the Welfare State 1942–74. Rivers Oram
5. Gibb AGF (1999) Off-site fabrication: prefabrication, pre-assembly and modularisation. Wiley
6. Landolfo R, Pali T, Bucciero B, Terracciano MT, Shakeel S, Macillo V, Iuorio O, Fiorino L (2019) Seismic response assessment of architectural non-structural LWS drywall components through experimental tests. J Constr Steel Res 162
7. Iuorio O. (2007) Colf-Formed Steel Housing. Pollack Periodica. 2–3:97–108
8. Iuorio O (2024) Methodology for improving manufacturing and assembly of lightweight prefab systems. TEMA, https://doi.org/10.30682/tema100022
9. Iuorio O, Kechidi S, Banks N (2021) Experimental investigation into the performance of cold-formed steel walls sheathed with OSB and cement-based panels. In: EUROSTEEL 2021
10. Bryden Wood Technology Limited (2023) Platform in practice. A lean approach to industrialized construction
11. Massbespoken brochure. https://www.massbespoke.com/documents/MassBespoke%E2%84%A2-Brochure-2017.pdf

Chapter 4
The Role of Prefab in the Context of Climate Emergency

Abstract Building performance, embodied carbon, responsible material use and circularity are the drivers for sustainable buildings. Digital technology is used in smart systems to support processes for environmentally friendly design, construction, and operation. This chapter analyses the balancing act of Modern Methods of Construction between innovation and climate emergency. It discusses the latest innovations embedded in prefab systems to deliver low-carbon systems, advance circularity in the built environment and retrofit the existing building stock. This section draws a methodology to assess the sustainability performance impact on inhabitants, to reflect upon the complexity of delivering indoor comfort in a changing climate. It reflects upon life-cycle analysis of prefab systems developed to be highly energy efficient and seismic resilient, while reducing carbon footprint. It opens up to the challenges of deconstruction and reuse, discussing the opportunity to reuse lightweight façade systems to shed light on the carbon benefits of closing the loop.

Keywords Net-zero · Design for deconstruction and reuse · Life-cycle analysis · Embodied carbon · Retrofit

4.1 The Climate Emergency: Problem Definition

The consumption of natural resources has started to be questioned since 1970 when Georgescu-Roegen started elaborating on the limits of natural resources. His and Howard Odum's theories were at the basis of what was then called the sustainable development of our society. In his book "Environment, Power and Society" Odum [1], in particular, stated that "the true value of energy to society is the net energy", which means there is a need to consider the one left, once the subtracted energy has been detracted. In the same years the energy crises, due to the rise of the cost of the oil exported by the Arabian countries, have accelerated the attention towards energy consumption patterns. Indeed, in those years, Rosenfeld noted that the American countries consumed twice the energy per capita, compared to European countries with the same life standards. Hence, consumers' behavior needed to be re-thought.

O. Iuorio, *From Mass Prefab to Mass Customization*,
PoliMI SpringerBriefs, https://doi.org/10.1007/978-3-031-73327-7_4

He started leading energy efficient studies, which brought to the definition of key building technical elements to be reconsidered: starting from the building envelope, to windows, lighting, HVAC system and zone design.

The first net zero building was built in 1976–1977 near Copenhagen, and with high insulation materials, solar collectors and water storage, posed the basis for the development of net zero in Europe.

In 1988, the Intergovernmental Panel on Climate Change (IPCC) was established to assess scientific information related to climate change and to provide a scientific basis for governments to develop climate policies. The increase of extreme weather events caused the IPCC to release a series of technical assessment reports (TAR), of which the third one, often referred to as the AR3, highlighted the urgent need for action. The TAR provided stronger evidence that the climate is warming, with global surface temperatures increasing by about 0.6 °C during the twentieth century. Future projections estimated an increase in global average surface temperature by 1.4 to 5.8 °C over the period 1990 to 2100, depending on various scenarios of greenhouse gas emissions. They predicted impacts on the ecosystems, with possible changes in flora and fauna, and severe impacts on human health, some of which are already visible today, with the increase of mortality due to severe urban heat and the arrival of dengue in Europe. Moreover, the increase of periods with either very low temperatures or extremely high temperatures and no precipitations are determining stress effects on both the built environment and the inhabitants.

At the building scale, the effect is that prolonged periods of high temperatures spaced out with low-temperature can cause materials to expand and contract, leading to cracks and structural damage; at the same time, increased temperatures for a prolonged period lead to higher demand for air conditioning, raising energy consumption and operational costs. This last is evident, not only in hot climates, but it is more and more evident in temperate and continental climates, where we are witnessing that buildings that were recently designed and manufactured to provide internal comfort with passive house strategies are suffering from high temperatures during the summer months, as they have limited capacity to cope with them.

Despite the international efforts to invest in passive house strategies to cope with cold and continental climates and to invest in net-zero policies, according to the International Energy Agency (IEA) the Carbon Dioxide (CO_2) emissions, produced by building operations and construction processes in 2021 were equivalent to 37% of global emissions [2]. Therefore, in this context, this chapter aims to analyze how some experiences of prefab have tried to tackle the **reduction of the operational impacts through net-zero developments, or reduce carbon impacts by lowering the embodied carbon, or enable reuse by implementing circular practices through either design for deconstruction and reuse (DfDR) or extending the lifetime of existing buildings through retrofit approaches**.

4.2 Development of MMC Systems to Achieve the Net Zero Targets

Many authors have written about net-zero housing technologies and promises, but not enough has been still discussed about its achieved performances. This section after reviewing international applications of low-carbon and net-zero prefab systems developed for new and existing buildings, it discusses methodologies for post-occupancy evaluations, and using as a case study a recent development in North of England analyzes the impact of prefab housing in the field of indoor comfort. This section aims to propose a multidisciplinary methodology that could be further adopted to evaluate the performance of prefabricated buildings.

4.2.1 Net Zero Solutions for New Construction and Retrofitting the Built Environment

The 2030 Agenda for Sustainable Development [3] has an objective to make cities and human settlements inclusive, safe, resilient, and sustainable. The 2020 Europe Strategy [4] and the 2050 Roadmap [5], regards radical improvements in energy efficiency and carbon emissions as ambitious and essential objectives for both new and existing buildings.

Net Zero houses are those that are able to generate as much energy as they need for heating, cooling, hot water, lighting, ventilation, and to power the appliances. A variety of measures have been developed over the last decades to increase energy efficiency of buildings and achieve net zero, including passive energy strategies, including high-performance fabrics to reduce the heating and cooling loads [6], and the integration of renewable energy sources, such as solar panels, heat recovery systems, and effective energy management systems.

However, the concept of net zero requires very careful considerations, as the goals of sustainable development, which include achieving a fair net-zero transition, socio-ecological sustainability, and the expansion of economic opportunities, must be in line with net zero [7].

In the prefab industry, we have witnessed two trends. The first one applying net zero energy concepts into the design of new housing, such as, for instance, in the case of BNZEB, a nearly zero energy building based on a cross-laminate timber (CLT) prefab system built in Mediterranean climate. CLT is finding large applications in current sustainable architecture as it is considered a green bio-based material, formed by wood and glues, which do not emit solvents or formaldehyde. CLT is implied in the BNEZB for both wall and floor panels and it is coupled to one or more layers of wood wool thermal insulation, with different density and thermal conductivity, respectively, to achieve the required thermal performances (description of the walls and roof is provided in Sect. 4.3.2). The building has been design as an "all-electric net zero-energy house". As energy source, the building uses an air-to-air heat pump to provide mechanical ventilation, both in winter and summer, and the system is also equipped with an active thermo-dynamic heat recovery [8]. Moreover, a large photovoltaic system is adopted as energy source and during the heating season, solar collectors are used to pre-heat the ventilation air during the day, while geothermal boreholes are used to pre-cool the ventilation air during the summer. This is just an example of how prefab systems can be designed and optimized to reach high-performance fabric and to be integrated with renewable energy sources, to achieve the net zero.

The second pathway that has seen the adoption of prefab systems for achieving net zero, has been in the retrofit of existing buildings. Reducing the energy consumption of existing buildings is, indeed, fundamental to meeting the European statutory 2050 decarbonization target. The 'Retrofit' term as defined by the ISO 6707 of European Standards is used to indicate those modifications that can generate improved conditions of the asset. Retrofitting existing buildings poses urgent challenges. Most of the buildings that will need to be zero carbon by 2050 are already built. The housing shortage that the European countries are facing requires the current building stock to be used to its highest potential. There is a need to extend the lifetime of current buildings according to different climate conditions and improved safety requirements. The request to intervene on a large part of the housing stock arises, in fact, from its manifest functional and technical-constructive inadequacy, i.e. from the high presence of widespread degradation phenomena, structural deficits, fruition and energy criticalities (with particularly high consumption both during the summer and winter period). Increasing energy efficiency of existing buildings is one of the biggest challenges for the construction sector. Retrofit of existing buildings can play an essential role in reducing carbon dioxide emissions and alleviating fuel poverty, reducing household energy bills and improving wellbeing when proper retrofit measures are developed [9, 10].

The major European experiment of retrofitting through the adoption of prefabricated systems has been developed under the Energiesprong initiative. The technological novelty of the Energiesprong initiative lied in the integration of digital scanning, BIM modelling, design for manufacturing (DfM), prefabrication of the customized exterior envelope and rooftop with solar panels that could improve the energy performance of the existing building, and assembly of the prefab customized panels in short time, with minimal disruption for the building users. Energiesprong also included the installation of smart heating and cooling systems to achieve the net zero. The initiative was born in the Netherland, and has found application also in France, Germany, UK and Italy. In Italy, specifically, the recent economic incentives have seen a proliferation of retrofit of multistorey, multi-owned buildings. This has been the opportunity for some industries to develop their prefab system for retrofit applications. These has been for instance the case for the development of the Irondom Duo system, as well as the application of some Leaphomes.

It is worth note here, the importance to adopt for any retrofit intervention an integrated approach. Indeed, the risks associated with approaching energy-related problems by focusing on a small number of straightforward characteristics (such as the envelope's thermal transmittance) can trivialize the issue and hinder the creation of suitable energy-performing solutions. For this reason, the criticalities of any simplified approach need to be taken into consideration, and instead an all-encompassing, integrated approach should be used to examine all the interconnected elements of a building's performance, such as its structural, energy, and aesthetic qualities, from a lifetime perspective. In fact, an integrated approach to retrofit can be weighed in accordance with its effects on the economy, society, and environment. The implementation of integrated solutions to multi-story concrete buildings around Europe is of great interest today, and was the main topic of the SURECON International Symposium [11]. The Duo system by Irondom is an example of a prefab lightweight panelized system, made out of a CFS frame, collaborating with OSB panel for the seismic action, and ultra-thin and heat-reflecting layers composed of atomised aluminized polyethylene, sheep wool or wadding and polyurethane foams for the energy upgrading [12].

4.2.2 *Methodologies for Assessing Performance and Sustainability of Low-Energy Buildings*

Gathering feedback from building inhabitants is crucial in understanding their experiences and satisfaction levels when they occupy buildings. Post-occupancy evaluation methodologies (POE) have been developed to assess various aspects of a building's performance after it has been occupied and in use [13]. POE provides valuable insights into how well the building meets the needs of its occupants, its environmental impact, and its overall effectiveness in achieving its intended goals. POE can assess how well buildings meet the needs of their occupants and how they function in accordance with the design intentions after they have been occupied and in use. When applied to prefabricated buildings, POE helps to understand the performance of these systems compared to traditional construction methods and can help identifing areas for improvement in future projects.

POE of prefabricated buildings involves evaluating their performance in key areas such as thermal comfort, indoor air quality, energy efficiency, acoustics, durability, and overall occupant satisfaction. Data on these aspects can be collected through occupant surveys, interviews, focus groups, indoor environmental quality (IEQ) physical measurements, energy consumption analysis, and building condition assessments. Therefore, a comprehensive POE methodology comprises of both qualitative and quantitative data gathering. Quantitative data are very important and can be evaluated by monitoring building spaces with an independent set of temperature sensors and, at least, relative humidity data loggers, over a year. A comprehensive POE should indeed, interpolate qualitative data gathered through surveys, interviews and focus groups, with quantitative data gathered through IEQ monitoring and weather/temperature data, collected over the same investigated post-occupancy period. However, methodologies for quantitative data gathering and analysis are out of the scope of this book. Readers interested in this topic can refer to Zimring [14], Westminster Guide [15] and Oseland [16].

When new data of post-occupancy are generated, these last are evaluated against existing databases. Among the most commonly used benchmarking POE databases, it is worth considering the Center for the Built Environment database, the Building Use Studies database [17], and the Building Occupants Survey System Australia database. In the following section, an application of a qualitative post-occupancy evaluation in a recent prefab development is analyzed to determine to what extent the project meets its promises of being an environmental and social sustainable place to live.

4.2.3 Performance Evaluation of a Low-Energy Prefab Development in Oceanic Climate

In an attempt to investigate the impacts of prefab on the occupants, and to understand if the promises of sustainable and comfortable housing are usually met, a recent study carried out at the University of Leeds and supported by the Politecnico di Milano, looked at six prefab case studies. It is remanded to the Prefab Sprouting project scientific reports for the full explanation and discussion. Here, only one case study, which has been under investigation since 2016 is presented and discussed.

An Innovation District, located in the North of England, can be considered an emblematic case study of recent prefab development, that aimed to transform a brownfield into a social housing scheme, with the articulation of 104 dwellings, distributed in terrace houses near the river (Fig. 4.1). The project included the entire urbanization of the area, including green spaces and a footbridge, and will comprise in the future also a school. The project developed primarily with private funding, has been since the beginning supported by the local council for the promises to provide environmental sustainable and affordable housing, in area of a city that was considered quite neglected despite the strong proximity to the center. The developed prefabricated system is based on wooden walls, filled with bio-insulation, lined internally with gypsum boards, and adopts colorful cement-based boards as exterior finishing. Most of the houses are articulated over three floors, having the bedrooms and a small office on the ground and first floor, the kitchen and living space on the second floor, and a roof terrace. The houses are equipped with triple glazing, mechanical ventilation, a heat recovery system, and an air source heat pump, to achieve low-energy standards.

For the production of the prefab system, an ad-hoc factory was built near the site on one side of the river, in which the wood coming from Poland was cut to size, and assembled in wall and floor panels, filled with the insulation, pre-treated for the integration of electrical and plumbing services, and completed with the finishing, to then be assembled vertically and horizontally in the nearby site, where also windows and openings and final finishing were completed. This specific organization of the

Fig. 4.1 Climate innovation district: panel production (images on the top), and the social housing development (bottom)

process, makes this case quite unique, as it can be classified as an MMC category 2, which couples mass-production with some degree of customization, delivered with a sort of flying factory [18], built on the site, together with an entire office building connected to it, for the design, commercialization and management of the entire project delivery. This structure, perhaps, slowed down the construction process, which has gone through multiple variations over the past four years.

Understanding the impact of this project on the inhabitants has been critical to verify if and how the sustainability promises were met. According to this last objective, a survey and a series of interviews were conducted. The post-occupancy evaluation survey was articulated around four main themes, which were related to: (1) the location of the housing development to understand the level of inclusion of the site within the city context and the access to green and local amenities; (2) the architecture, to understand the quality of the spaces, sizes, and the materiality of the indoor environment; (3) the indoor comfort, during both winter and summer months; (4) the flexibility and adaptability of the houses to future needs.

Looking closely at the indoor comfort, the temperature and air quality were investigated. Specifically, it was found that in winter (Fig. 4.2), the indoor temperature was on average set around 19,5 °C, and the house felt quite warm, with little variation throughout the day, therefore, overall on a scale from 1 to 7, the indoor winter temperature resulted comfortable (6 out 7). Similarly, the quality of the indoor air during winter months resulted to be quite good, and be dry, still and odourless, although in some rooms it can felt slightly stuffy (4 out of 7). Looking instead to the summer months, the situation is clearly different. Indeed, Fig. 4.3 shows that both the indoor temperature and the indoor quality in summer are overall uncomfortable, reaching very high indoor temperature, and high levels of humidity (an average value of 2.4 out of 7, where 0 indicates extreme humidity), with air being very smelly and stuffy (with an average value of 2.3 for being smelly and 3 for being stuffy). These results demonstrate that while the houses have been designed to be very energy efficient in winter, with good envelope quality, efficient air tightness, and triple glazing, the same have not being designed and executed to cope with the summer months. This reflects the tendency of giving priority to solve the indoor comfort for the low temperatures, which in the area, goes from 5 to 15 °C for the majority of the year. However, the increasing of hot temperatures in some weeks over spring and summer, make inhabitants to struggle to cope with summer temperatures, as the houses lack a strategy for natural ventilation and or dehumidification, without having proper openings or cooling devices. Moreover, the presence of large openings, that fuel the indoors with natural light, which is greatly appreciated throughout the year, at the same time lacking of any shading device, makes it difficult to reduce solar gains, on the hot days.

Therefore, this case study is considered a very good example of prefab system that while delivering its promises of providing a sustainable and energy-efficient solution

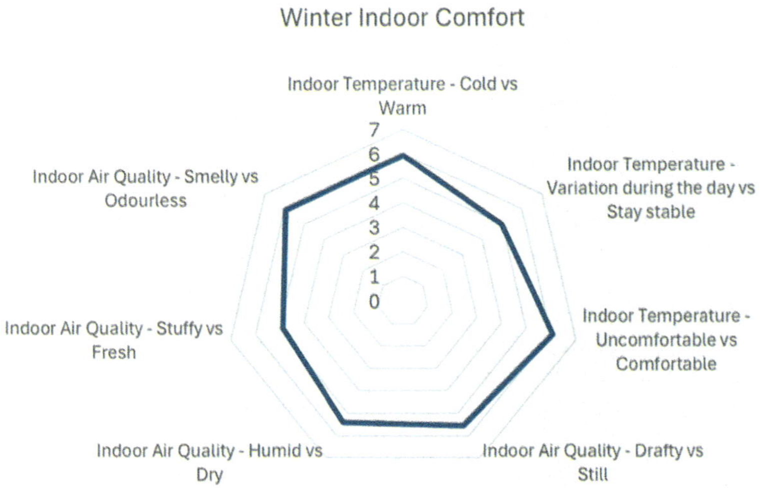

Fig. 4.2 Winter indoor comfort as perceived by the occupants in terms of indoor air quality, and temperature, with each parameter is evaluated on a 1 to 7 scale

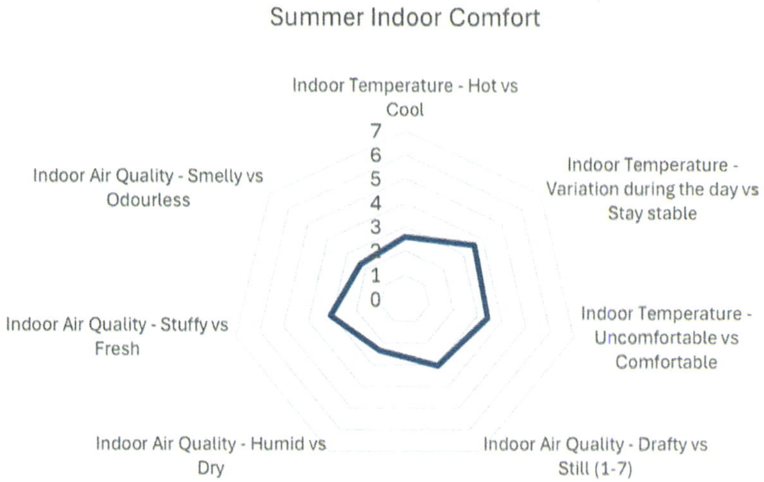

Fig. 4.3 Summer indoor comfort as perceived by the occupants in terms of indoor air quality, and temperature, with each parameter is evaluated on a 1 to 7 scale

in winter, it also shows the limitations of the very efficient envelope to cope with climate change, which is bringing the recurrence of extremely hot periods.

4.3 Low-Embodied Carbon Potentials in a Life Cycle Perspective

A number of European laws and policies have been recently developed to promote and assist the building industry's environmental sustainability. They encourage using life cycle assessment (LCA), which comprises operational and embodied energy (EE), since the beginning of the design process, to critically assess and optimize material and energy use. They also provide guidance for adopting building systems, procedures, and materials that have less of an adverse effect on the environment, produce less waste, and use less water. There is a tendency toward reducing Embodied Carbon (EC) emissions in projects and requiring the reporting of carbon emissions in the built environment. EC is defined as the total carbon emissions, calculated in terms of CO_{2e}, associated with the extraction, production, transportation, installation, use, refurbishment, replacement and disposal at the end of life of products, but excluding the carbon used for operation. EC can be evaluated by following detailed LCA according to ISO 14040/44 [19], or by adopting simplified procedures, that can estimate the EC by multiplying the mass of materials for their embodied carbon coefficient, where this last can be either retrieved by existing databases or by the Environmental Product Declarations (EPDs).

4.3.1 Why Embodied Carbon Matters

The importance of evaluating and reducing EC in buildings has become prominent since the construction industry started to properly reduce the energy used to operate buildings. Indeed, operational energy typically has a larger impact on the total life cycle energy of a building over long lifetimes. However, as the operational efficiency of new buildings is improved, the relative significance of the embodied impacts of construction materials and processes increases. Moreover, the request for building fabrics with higher thermal performances, to reduce the energy used for heating, can come at the expense of using a larger volume of materials, which can, unfortunately, have higher carbon impacts.

Therefore, there is a delicate balance between operational energy and embodied energy that needs to be considered for the achievement of sustainable construction systems. In the following sections, after highlighting why EC matters, the comparison of EC for two MMC systems and a traditional Southern Italian construction are discussed, and aspects that could be considered in the future to further reduce EC of carbon intensive systems are presented.

4.3.2 Evaluation of Embodied Carbon in Wooden Based and Steel Based Prefab Systems

In recent years, many researchers and industries have attempted to understand what are the construction systems with lower embodied carbon. The truth is that there is not a definitive answer. While, certainly, those systems using a high mass of material per square meter, and having low recyclability and reusing potential, tend to have high EC, when lightweight prefab systems are under investigation, it is difficult to say what system in absolute is to be privileged. This is certainly one of the conclusions, that was drawn by the Author research team [20] when a life-cycle comparison was developed between a cold-formed steel school designed and built in Naples, which is representative of a category 3 of MMC, a cross-laminated panelised system (category 2 of MMC) used for a net zero house in Benevento (BNZEB in Italy), and a typical masonry construction.

Specifically, the investigated CFS MMC3 system was that employed for a British School (BFS) in Italy in 2009 (Fig. 4.4), and it comprises six interconnected single-story stick-built CFS buildings, each approximately 4 m in height. The school's design adheres to the Italian OPCM 2006 standards for a grade 2 seismic zone and was executed to achieve optimal energy efficiency. All walls and roofs are constructed using CFS components, employing sheathing-braced methodologies that account for the interaction between steel profiles and sheathing panels to attain the requested racking capacity [21]. The steel elements, including wall studs, tracks, rafters, roof joists, and blockings, are fabricated from zinc-coated, hot-dip galvanized steel grade S320, with thicknesses ranging from 1.5 to 3.0 mm. Structural sheathing panels consist of 9 mm thick type 3 Oriented Strand Boards (OSB). The building's functional and technological "packets" for walls, floor and roof were designed to enhance eco-efficiency throughout its lifecycle. These include the incorporation of wood wool panels (CELENIT) and hemp fiber as thermal insulation materials for walls and roofs, as well as gypsum-based panels for fire resistance. Detailed sections of the walls and floors are discussed extensively in Iuorio et al.20.

The BNZEB is a detached home in Benevento, Southern Italy, situated in the same climate zone of the BFS, and built for a seismic zone 1, which is defined by a very high seismic risk [8]. The CLT walls are made up by CLT panels with five crossed layers of class C24 solid wood boards, each panel measures 9.5 cm in thickness (Fig. 4.5). The roof is made up with the same typology of panels, but a thickness of 12 cm. Wood wool thermal insulation layers are adopted for walls and roof to achieve the required thermal transmittance (i.e. 0.17 W/m^2 K for the walls and 0.23 W/m^2 K for the roof).

A masonry construction with perforated bricks of 420 mm thickness, filled with reinforced concrete, and lined with 70 mm of mineral wool, was considered for comparison, with the previous two prefab systems, as representative of a typical construction in Southern Italy, and having a thermal transmittance of 0.22 W/m^2 K (Fig. 4.5). For the roof, instead, a traditional flat roof made of concrete blocks, with 100 mm of expanded polystyrene (EPS) was considered to achieve a thermal transmittance of 0.28 W/m^2K.

Given these three systems, an LCA in a cradle-to-gate perspective, which includes only the environmental impacts associated with raw material production (A1), transport (A2), and manufacturing (A3) was developed according to the current standards. To compare the three system typologies, 1 m^2 of opaque wall and 1 m^2 of roof were considered as functional units. The inventory (LCI) was developed according to the EPDs, when known, or retrieved from the One Click database [22].

The comparison showed that for the external walls, the EC of the CFS system has the highest value with 123 $kgCO_{2e}/m^2$, and this value is very close to the traditional system, with EC of 108 $kgCO_{2e}/m^2$. In these wall configurations the main contributor to EC are for the CFS walls, the steel profiles which account for 51% of the carbon impacts, and for the traditional wall the clay bricks which are responsible for 65% of the impact.

The comparison of the three square meters of roofs, shows the highest EC for the traditional roof system, with 98 $kgCO_{2e}/m^2$, which is about 15% above the EC of the CFS roof and 53% higher that the CLT roof. The configuration envisaged for the CFS roof achieves a level of thermal insulation comparable to that obtained in the case of the CLT, with a total thickness reduced by more than 50% but with a carbon footprint higher by 32.8%.

Overall the prefab system based on the use of CLT appears to be the one having the best carbon performance, thanks to the use of bio-based materials. Moreover, the study also demonstrated the importance of optimizing the structural systems, as any reduction in the use of steel, for the CFS stick-built solution and of brick for the traditional building, would result in a reduction in EC. The study, detailed in Iuorio et al. [20], also develops a detailed LCA accounting for all the seven greenhouse gas

Fig. 4.4 Cold-formed steel school built in Naples in 2009 and named BFS, showing the construction process together with the final building (photos by the Author)

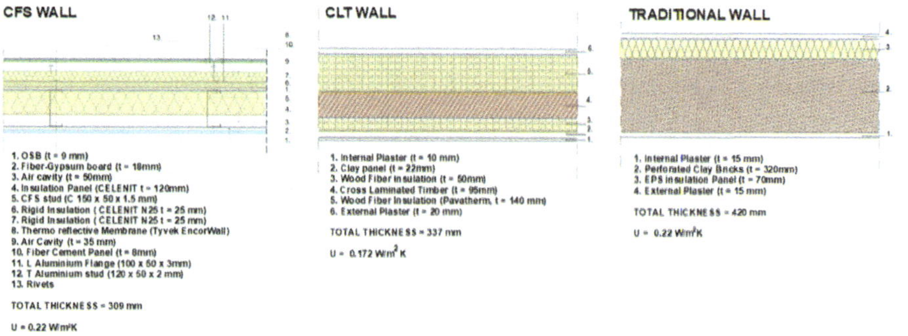

Fig. 4.5 Wall stratifications for a lightweight steel system (left), a CLT net-zero building (center), a traditional building in Southern Italy (right)

emissions, in the cradle-to-gate boundary, and clearly, demonstrates that the impacts are primarily connected to the material quantities, rather than the prefabrication level. Although, this last can play an important role in reducing the waste generated during the construction process.

4.3.3 Future Opportunities to Reduce Embodied Carbon

The previous section shows that even lightweight systems can be carbon intensive, when they are based on steel use, and are developed for achieving high thermal and structural performance. This acknowledgment should not be translated into avoiding their use, while instead should be considered to critically consider how further reduce their impacts. Given that impacts are associated with the mass of materials, their transport, manufacturing and the carbon factors, efforts should be directed to continue to improve each of these factors.

In terms of mass of materials, its reduction should be done not only by optimizing the system and reducing the overengineering, but also, for example, by employing higher-strength steel, that could allow for the use of thinner cross-sections, and by improving their reusability, so that rather than counting on the recycling of the steel, the components could be reused over multiple cycles, and hence obtaining carbon discount.

As per the transportation and construction, the carbon reduction can be connected to their decarbonization. For instance, according to the UK Green Building Council [23], there could be a reduction of more than 15% of carbon in transportation by 2030, due to the phase-out of new petrol and diesel vehicles for road freight, which is the predominant mode utilized in construction, and a reduction up to 50% by 2050 with the use only of zero-emission vehicles. UKCBG also reports a decrease in construction site emissions of 5% every year due to the optimization of the construction processes, digitalization of construction operations and processes, value engineering, and behavioral training of the operators. Projecting this reduction in the future would bring a significant reduction of EC connected to manufacturing and construction, and this reduction can be even more evident in factory-based manufacturing, such as per MMC.

Finally, the decarbonization of the foundation industries, i.e. steel, cement and concrete, glass, bricks, plastics, and timber can strongly reduce the impact of construction. This would require ingent nationalized actions.

4.3.4 Circular Systems: Can MMC Enable Circularity?

The current linear construction approach of "take-make-use-waste" necessitates enormous natural resource extraction, and generates a large amount of waste and carbon emissions, to the point that the construction industry is considered responsible for 40% of the world carbon emission, and construction and demolition waste (CDW) accounts for 36% of European waste [24]. The world's population will increase exponentially from 7.5 billion to about 8.5 billion by 2030, adding to the already high demand for housing and services and putting unprecedented strain on natural resources. Construction activities, such as extracting resources, producing building elements, and transporting them across distances result in resource depletion and global warming of the Earth's environment. Moreover, the current difficulty to access some of the resources due to global instability and many countries' unrest, as well as supply chain disruptions that can be caused by pandemics, such the recent COVID19, are causing uncertainty in the short term and increasing overall prices.

Circular economy principles have the aim to break the linear approach and to bring back materials and systems in a continuous loop. Enabling circularity of construction would support the Waste Recovery Framework Directive, that requested European countries to recover 70% of their waste [25]. The recovery can be enabled through the treatment of nonhazardous CDW materials for retaining, reusing, and recycling. These last, correspond to the 3R principles of "Reduce, Reuse and Recycle", which are the most agreed-upon principles for circular economy in the construction industry and have also been articulated in the 10Rs of "Refuse, Rethink, Reduce; Reuse, Repair, Refurbish, Remanufacture, Repurpose; Recycle, Recover". In particular, these principles can be read as inviting, above all, the construction industry to Refuse to build new construction, unless strictly necessary, and to Rethink the construction processes to Reduce the amount of material used. When necessary, they invite to Reuse buildings, through, for instance, Repairing and/or Refurbish them to allow further use, or by Deconstruction and Reuse (DfR), or Deconstruction and Remanufacturing of building components. Finally, when none of the previous is possible, the CE principles request to Recycle materials, which means reprocessing materials to be then used in a new product. Smart application of CE principles would prioritize recycling through the most appropriate stream, if possible upcycling them rather than downcycling, in order to Recover their value [26]. It is important to note that, the order in which the 10Rs are defined reflects the hierarchy that should be considered to maximize the use of existing resources while retaining their value, over multiple lifecycles, to minimize their depletion.

This section after delineating the main circularity streams within the prefab industry, it discusses experimental and life cycle studies recently developed to support the circularity of lightweight prefab façades. Enabling the circularity of these last is of particular importance, as façade systems have usually a lifetime much shorter than the other components of office, school, and commercial buildings. Therefore, enabling their Reuse can have significant carbon impacts.

4.3.5 Circularity Streams Within the Prefab Industry

A critical advantage of MMC lies in the importance given to the optimization of resources, which is usually at the basis of the current MMC systems, as already discussed in Chaps. 2 and 3 of this book. Indeed, as largely demonstrated, Reducing the amount of material used for the construction of prefab systems is a priority for the current industry, as that helps in achieving sustainability targets while reducing costs. Therefore, from this perspective, it could be claimed that prefab tends to align with the first principle of CE.

Moreover, enabling the Reuse of prefab modules and/or modular components can play a pivotal role in minimizing environmental impact. Indeed, in theory, by prioritizing the reusing of building components, prefab can drastically reduce the demand for the production of new raw materials, leading to a substantial decrease in embodied emissions. This approach could not only preserve resources but also extend the lifecycle of existing materials, reinforcing a circular economy within the construction sector. Indeed, the inherent flexibility of prefab systems can be tailored to enable the adaptive reuse of modules for multiple lifecycles and maximize their potential technical value which is crucial for reducing waste and conserving energy.

By repurposing and refurbishing existing structures, through the adoption of prefab systems for the retrofit of existing buildings, prefab can also significantly cut down on the resources and energy required for new construction. This practice is instrumental in curbing GHG emissions associated with various stages of construction, from material extraction to transportation and assembly.

Therefore, emphasizing reuse and retrofit within prefab not only aligns with sustainable practices but also represents a transformative approach to reducing the environmental footprint of the construction industry.

An example of the application of circular principles in the prefab industry is provided by the Polyblock System, which could be categorized as an MMC6 system. The Polyblock is a prefabricated construction component designed as an alternative to traditional concrete and developed by Polycare Research Technology GmbH in Germany [27]. It is created with a focus on reusing waste materials and responsibly utilizing local resources. The block consists of two main parts: a cover or shell, and an inner filling made from EPS or mineral wool. The shell is made of polymer concrete, which includes 12% unsaturated terephthalic polyester resin, incorporating up to 38% recycled PET as a binder and up to 88% filler material, such as local sand or secondary raw materials like slag, tailings, or construction and demolition waste. The Polyblock offers excellent mechanical strength, outperforming cement concrete in both compression (90–130 MPa) and bending. Additionally, the block's design provides enhanced thermal insulation, with wall thermal transmittance values ranging from 0.4 to 0.55 W/m^2·K. Despite its high performance, the block remains lightweight [28]. The Polyblocks are assembled together by stacking them up, and they are connected with threatened rods, without using any cement or mortar. This also facilitates further deconstruction.

The Nexii Discovery Centre, constructed as a demonstration project exemplifies, instead, the principles of Design for Deconstruction (DfD) embedded in a prefab system belonging to the category 2 of MMC. Initially manufactured in Nexii's Moose Jaw plant and assembled in Squamish in 2019, was disassembled in only 6 days in 2021 applying a construction reverse logic, and is planned for reassembly as a laneway house on Saltspring Island. This project serves as a case study in circularity, showcasing the potential for building panelized systems to be reused at the end of their service life. Indeed, in the deconstruction process, only 0,06% of the material went ot landfill and corresponded to 2 buckets of mixed waste [29]. Nexii's approach to DfD emphasizes a panelized construction system designed for ease of disassembly and reassembly. Key features of this system include the use of durable materials such as Nexiite, modular wall assemblies for simplified connection and disconnection, accessible bolt connections, and the integration of mechanical, electrical, and plumbing systems with a focus on future maintenance and renewal cycles. Additionally, panels are labeled and cataloged (with labels including the type of panel, size and weight) to ensure that essential information is available for future use. The Discovery Centre project highlights both the advantages and challenges of implementing DfD, offering valuable insights into the design and engineering considerations necessary for achieving true circularity in construction.

4.3.6 Circularity of Cold-Formed Steel Façade Wall Systems

Lightweight façades are MMC type 5 systems, that are largely used as exterior walls for concrete and steel frame buildings, from single-storey to multy-storey residential, office and commercial buildings. They are widespread thanks to their extreme versatility in composing façades at a fast pace, satisfying nearly any opening requirement, and allowing composition with a wide range of thermal and acoustic insulation to achieve the required performance. Moreover, the possibility to tune the assembly process according to each project requirement and the lightness of the components make them ideal for speeding up construction processes, as these façade systems can be assembled by bringing on-site elemental components (cold-formed steel (CFS) studs, tracks, blocking, and sheathing panels) that are connected with fast fastening on site, or sub-wall assemblies can be pre-composed in factory to be then erected on site, either by hand or with small cranes. The interface between the

CFS façade and the structural frame can either be obtained by connecting the wall tracks to the frames through screws. In case of high wind or seismic loads, stiffer connecting plates (stiffeners or hold-downs) or angle profiles can be adopted. Even when the primary structural frame shows built-up geometrical inaccuracies, the CFS façade systems can be flexible enough to be adapted to them through either the tuning of the length of the studs or through the composition of C and U-shaped profiles in box configurations or with the adoption of omega profiles. In any of these situations, the CFS wall façades are designed and fabricated to carry their self-load, the façade finishing, and the wind loads.

The simplicity of their erection process is also key to the ease of demolition, and, indeed, this system typifies the linear construction practices of "take-make-dispose". Figure 4.6 indeed shows the end-of-life of typical CFS profiles of façade and interior walls during building renovation processes, and it demonstrates that the easy handling of such components makes them prone to be taken apart and deformed to be landfilled, despite their residual bearing capacity. In an attempt to demonstrate their possible second life, a few of the profiles in Fig. 4.4 were used in the summer workshop "Whispers from a Summer Nest. A Circular Pergola from Reclaimed Materials" at the Politecnico di Milano, for the built-up of outdoor seating arrangements. Certainly, an application of this kind corresponds to downcycling the profiles from wall bearing to furniture, but, regardless it shows the possibility that a more careful use of such energy-intensive components could have if better managed at the end of their lives.

Enabling the circularity of lightweight façade systems has been one of the objectives of the Author research group within the UK-funded circular economy center "UKRI Interdisciplinary Circular Economy Centre For Mineral-based Construction Materials". The work, specifically, looked at investigating the possibility of disassembling and reusing lightweight façade wall systems made of CFS profiles and sheathed with gypsum and/or cement-based panels, by: (1) experimentally testing the structural behaviour of façade system after being disassembled and re-assembled; (2) by testing the opportunity of adopting bolted connections instead of screwed connections between tracks and the main building structural frame; (3) by quantifying numerically the environmental impact of multiple life cycles of the system.

In the first phase, two CFS lightweight skeletons, representative of lightweight infill wall systems in UK, were subject to a cycle of assembly, disassembly and reassembly. The observations made during the disassembly and reconstruction procedures demonstrated that the sample CFS frame was reassembled without causing any difficulty and that the screws connecting the CFS profiles, when carefully removed, could have been reused. The scientists also found that there was minimal difference in the lateral, torsional and out-of-plane stiffnesses of the frames between their pre- and post-deconstruction measurements. Based on this observation, and the notion that connection behaviour is critical to develop demountable systems, as largely investigated in Kitayama and Iuorio [30], it was decided to further investigate the possibility of reusing screw and CFS profiles.

The connections between infill walls and the primary structural frame experience shear loads under serviceability conditions. Therefore, 21 single lap shear tests between CFS and hot rolled (HRS) plates, connected by Hex Head self-drill screws

Fig. 4.6 CFS façade and partition steel components: left, typical end-of-life; right, reuse of some of the profile for making outdoor furniture

Thickness CFS t_1 (mm)	Thickness HRS t_2 (mm)	Connection	Number of test
1.6	15	Screw → Test	3
1.6	15	Screw → Unscrew → Screw → Test	3
1.6	15	Screw → Unscrew → Screw → Unscrew → Screw → Test	3
1.6	10	Screw → Test	3
1.6	10	Screw → Unscrew → Screw → Test	3
2.0	15	Screw → Test	3
2.0	15	Screw → Unscrew → Screw → Test	3

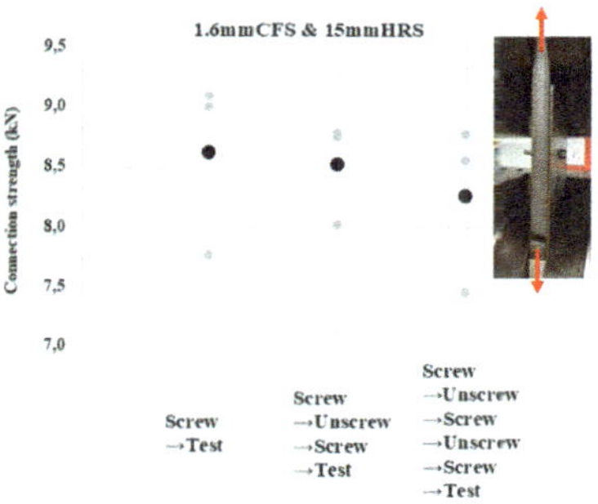

Fig. 4.7 Experimental testing of screwed connections between CFS infill wall façade and primary hot rolled structural frame: left, test program; right, shear test results

with 5.5 mm diameter, were performed. These were defined to replicate the current connecting system used between CFS façade system and primary structural frame (when this last is made by traditional steel frame). The tests were planned to simulate, first the building and using of the system, and then either 1 or 2 cycles of deconstruction and reuse. The results (Fig. 4.7) indicated that the unscrew and re-screw (i.e. each cycle of deconstruction and reuse) may cause a slight reduction of the strength of the connection, which is, despite all, nearly negligible at the first cycle of reuse, and can start to become evident only in a second cycle. Full description of the experimental study can be retrieved in [31]. Therefore, the tests reistate that mounting and demounting of the lightweight steel façade system does not affect the wall system bearing performances, and that could be reused at least once, in the same configuration with no problem, while for follow up reusing cycles, remediations, such as for instances, placing the connection in a new hole, or using a new screw, could be considered.

In an attempt, to make more robust the disassembly and reusing of lightweight CFS infill façade, the Author team investigated the possibility of using bolted connections instead of screwed connections, between the CFS tracks and the primary frame [32]. Therefore, an experimental campaign was carried out to investigate the behaviour of bolted connections, adopting as variables the CFS plate thickness (1.2, 1.6 and 2.0 mm), CFS plate width (50 and 75 mm), HRS plate thickness (10 or 15 mm) and the presence or not of the washer. A grade 8.8 bolt with a 16 mm diameter was used for the campaign. A total of eighty-two single lap shear tests were performed (Fig. 4.8). The force deformation curves showed and elastic–plastic behaviour up to the bolt head bearing in the thin plate, then tearing of the thin plate occurred. After the test, it was possible to see that while the CFS plate was broken, both the HRS plate and the bolts had no deformation. Therefore, both of them could have been used multiple times. The test results demonstrated that the use of a washer increased the peak strength of connections, while the thickness of the hot-rolled steel plates (10 or 15 mm) did not affect the strength of the connection, as the thin CFS plate governed the failure mode and strength. When we compare the peak strength of the

5.5 mm screwed connection between 1.6 mm CFS plate and the 15 mm HRS plate, with the 8.8 grade and 16 mm bolted connection between 1.6 mm CFS plate and 15 mm HRS plate, this second one is between 2.5 and 3.3 times higher than the screwed connection.

What are then, the overall considerations that can be made? Lightweight steel infill wall systems are stratified constructions made of CFS studs and tracks that can be safely disassembled through unscrewing of elements, and then reassembled in the same configuration without altering the bearing capacity. The connection between the CFS walls and the primary structural frames, can be made by screw or bolt connections. Screws are the most used, as they are cheaper, easier to be installed, and require access only on one side of the structural system. The experimental campaign showed that they can be unscrewed and reassembled to be reused once, without damage, while if reused more than once, they can start showing decreased capacity. Adopting instead bolted connections could allow the infill walls to withstand shear forces more than 2.5 times higher, hence they can certainly be adopted when very high shear capacity may be required. It also shown that even in the cases in which the track could reach the shear failure, in theory, the other wall members could be disassembled and reused with the installation of new tracks, while reusing the bolts and with no damage to the primary frame. Of course, this would require further testing and scenario analysis. Finally, clearly, the infill wall is more complex than just the CFS steel frame itself, as indeed this last is part of a façade wall stratigraphy that will depend on the aesthetic, thermal, and acoustic requirements. Therefore, the possibility of disassembling each of the components is all to be verified and considered. In the best scenario, when wall façades are pre-assembled in finished wall elements and access to the connections between the CFS and the floor or beam system is allowed, then the lightweight façade could be easily disassembled and reused.

4.3.7 Evaluating the Lifecycle Impact of Circular Lightweight Prefab Façade Systems

Understanding what would be the carbon impact of enabling the circularity of lightweight infill wall systems is necessary to evaluate the benefits and the potential challenges in moving towards circular construction systems. In an attempt to quantify the carbon emission associated with lightweight CFS infill walls, and to understand if any carbon reduction would be associated with making them circular, a detailed life-cycle study has been carried out, using as a case study a school building in North of England [33].

The investigated school comprises a three-floor HR steel building with concrete slabs, and concrete foundations. The infill walls have CFS studs and tracks, sheathed internally with two 12.5 mm gypsum boards, and externally by one 12.5 mm gypsum board, to achieve the required fire performance. The thermal insulation is provided

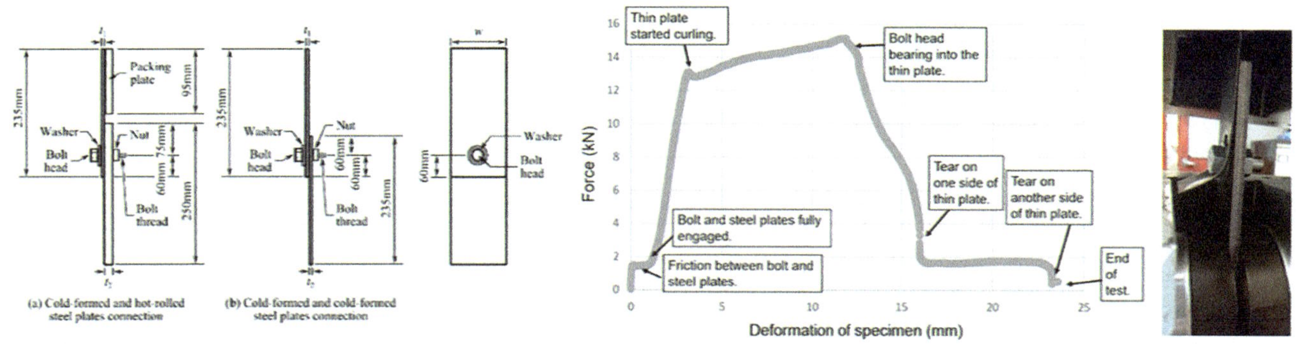

Fig. 4.8 Bolted connection testing for the design for deconstruction and reuse (DfDR) over multiple cycles of CFS façade systems: left, geometry of the tested specimens; center, typical force–displacement curve of a bolted connection between CFS and HRS plate; right, typical failure mode with the bearing of the CFS plate

with 200 mm glass mineral wool located in between the CFS studs, and by 100 mm Rockwool towards the exterior of the walls. Aluminum cladding completes the school façade.

The LCA study was carried out according to the ISO 14040 and ISO 14014. The goal of the LCA was twofold: it aimed to estimate the carbon impact of the façade system in comparison to the full building when the current linear scenario of the "take, make, dispose" is considered; and then to compare the carbon impacts of the linear scenario with those evaluated in a circular scenario case. Specifically, the LCAs were performed considering system boundaries in a cradle-to-grave approach, evaluating embodied carbon (EC) associated with the extraction and manufacturing of the building components, transportation to the site and installation (A1–A5), EC connected to the replacement of the infill façade system after 30 years (B4) and end of life, including recycling and landfilling (C2–C4). However, while in the case of the linear scenario, it was assumed that after 30 years from the initial construction, the infill walls were to be replaced by new infill walls (Fig. 4.9), to respond to changing energy and aesthetic requirements, in the case of the circular scenario, instead, the components of the infill walls having life expectance longer than 30 years were identified and it was hypothesized they would be re-used again in the new facade configuration (Figs. 4.9 and 4.10).

For the LCA, an inventory of all the masses of components of the structure, foundation, and lightweight facade system of the case study building was computed, and for the embodied carbon factors (EC, calculated in the form of CO_{2e}), the Environmental Product Declarations (EPDs) and the database included in RICS [34], the ICE documents [35] were considered. Table 4.1 summarises the masses and EC coefficient adopted in the study [33]. The embodied carbon impacts have been evaluated multiplying the mass of each component for the EC factors shown in Table 4.1.

The results of the LCA showed that the entire building has an impact of 196 $kgCO_{2e}/m^2$ when considering only the module A1-A5, i.e. according to a cradle-to-gate analysis. This value, overall is relatively low, given that it is considered typical for similar construction to have an EC ranging between 150 e 400 $kgCO_{2e}/m^2$. When the cradle-to-grave approach is evaluated, then the EC rises to 1294

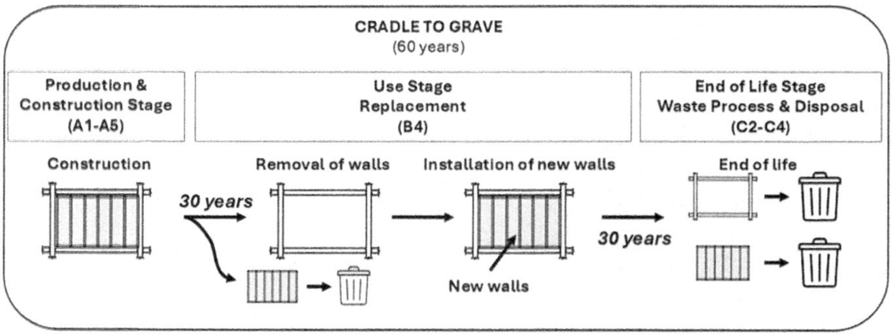

Fig. 4.9 Linear scenario for lightweight CFS infill walls

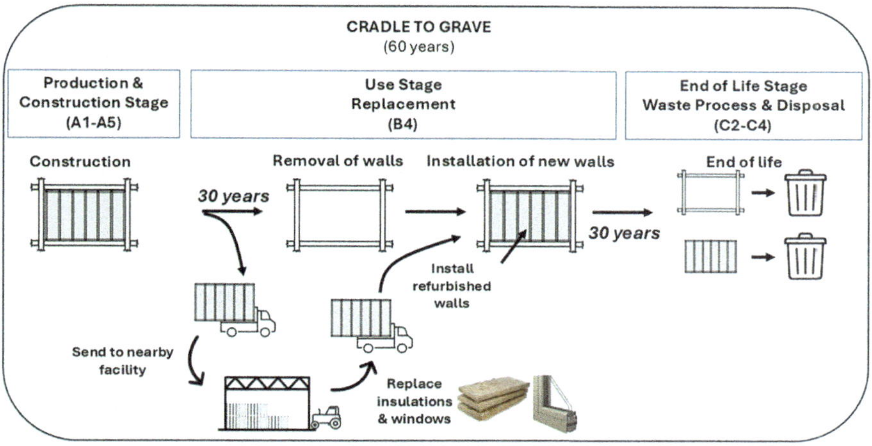

Fig. 4.10 Circular scenario for lightweight CFS infill walls

tonnesCO_{2e}, which corresponds to 225 kgCO_{2e}/m^2 for the linear scenario, and to 1219 tonnesCO_{2e}, which corresponds to 212 kgCO_{2e}/m^2 for the circular scenario. Of those EC the lightweight façade contributes to 22% of the EC impact of the entire building in the linear scenario, and that contribution is reduced to 17% when the circular scenario is achieved. Therefore, the study demonstrates that the lightweight façade system has a significant impact in the embodied carbon impact of an entire building, given that it account in between a fifth and a fourth of it. The study also shows that enabling the circularity of the system can save 6% of the overall EC.

The study also demonstrates that when the linear scenario is considered, with the reusing of all the lightweight façade components, except for the windows (frame and glass) and the insulation, as it is considered that both of them will need to be replaced to conform with any new energy performance requirement after 30 years, then the EC impact of the energy-intensive components, i.e. CFS and aluminum will be reduced by 6% and 3%, respectively (Fig. 4.11). It is important to note that this study only considered a circular scenario within the 60-year lifespan of the building. Further benefits could be achieved by evaluating multiple reuse of the façade components, to further capitalize on the residual life expectancy of the components.

Table 4.1 Life cycle inventory with the indication of the case-study building's structure, foundation and lightweight façade system, their mass, and the embodied carbon coefficients considered for the extraction and manufacturing (A1–A3), the construction (A4) and installation (A5), the replacement after 30 years (B4), the transportation to the landfill (C2), waste processing (C3) and disposal (C4). For info about how every coefficient has been evaluated, please, refer to Kitayama et al. [33]

	Component	Mass	Embodied carbon coefficients					
		(kg)	A1–A3	A4	A5	B4	C2	C3–C4
Structure	Columns	63,076	2.5	0.005	0.025	0	0.005	0.013
	Beam	149,549	2.5	0.005	0.025	0	0.005	0.013
	Slab (concrete)	2,106,908	50.2 kg CO_{2e}/m^2	0.005	0.502 kgCO_{2e}/m^2	0	0.005	0.013
	Slab reinforcement	99,038		0.005		0	0.005	0.013
Foundation Lightweight CFS façade system	Foundation (concrete)	40,954	0.12	0.005	0.008	0	0.005	0.013
	Foundation (steel rebars)	1507	0.76	0.005	0.041	0	0.005	0.013
	Ground bearing slab (concrete)	663,242	0.12	0.005	0.008	0	0.005	0.013
	Ground bearing slab (steel rebars)	31,177	0.76	0.005	0.041	0	0.005	0.013
Lightweight CFS façade system	CFS members	21,991	1.29	0.057	0.028	1.39	0.005	0.026
	Internal Plasterboards	24,672	0.16	0.032	0.011	0.22	0.005	0.013
	External Plasterboard	16,827	0.14	0.032	0.0044	0.18	0.005	0.013
	Glass Mineral Wool—insulation	9252	1.28	0.032	0.235	1.56	0.005	0.013
	Rockwool—insulation	7710	1.28	0.032	0.235	1.56	0.005	0.013
	Aluminium cladding	2503	6.83	0.032	0.069	6.95	0.005	0.013
	Fiber cement cladding	10,134	0.57	0.032	0.019	0.64	0.005	0.013

(continued)

Table 4.1 (continued)

	Component	Mass	Embodied carbon coefficients					
		(kg)	A1–A3	A4	A5	B4	C2	C3–C4
	Window glass	17,467	1.67	0.032	0.091	1.81	0.005	0.013
	Window aluminium frame	2615	6.58	0.032	0.349	6.98	0.005	0.013

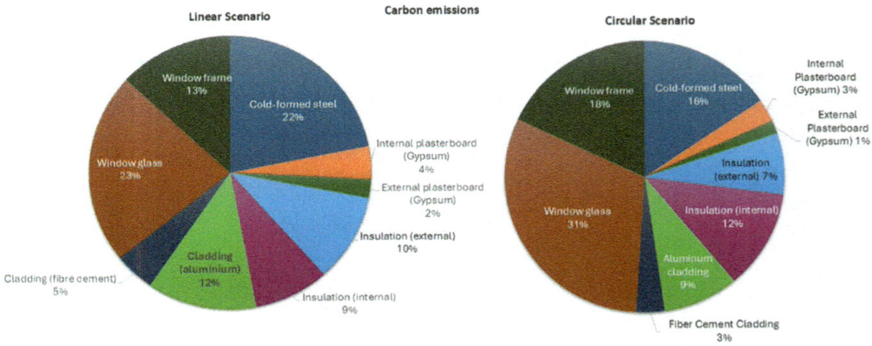

Fig. 4.11 LCA of the lightweight façades for the case-study building with the evaluation of the percentage contribution of each component in the linear scenario (left) and circular scenario (right)

4.4 Conclusion

The climate emergency is requiring designers, politicians, industry and users to act now to reduce the environmental impacts of our built environment, by increasing its energy efficiency, reducing waste, using at the highest value the materials, while at the same time prioritize the long-term wellbeing of the inhabitants.

This chapter after reviewing the definition of climate emergency and how its knowledge has challenged the construction industry, at a growing degree, for the last fourty years, it introduces to the concept of net zero buildings. It presents how prefabrication is playing a role in the leaning towards low energy and net zero for both new construction and the retrofit of existing buildings. In the retrofit paradigma, this book poses the attention in always considering integrated approaches, so that the retrofit action can be considered as an opportunity for improvement of aesthetic, energy and structural performance of the deteriorating built environment of our cities.

This work poses, then the accent on the importance to review the promises of energy efficient and sustainable prefab buildings through post-occupancy evaluations. An emblematic case of a sustainable development in Oceanic climate is cited, in which the energy efficiency promised and reached for winter periods, falls short in summer, as this can be nearly a new phenomenon that requires climate change adaptation to ensure climate resilience. This section demonstrates the importance of developing methodologies for capturing users' occupancy evaluation, as a means for understanding real performances, benchmarking indicators, and tuning systems to target the long-term well-being of the occupants.

Any consideration about climate adaptation of the built environment should look at the buildings in their lifetime. When we look at buildings in their lifecycle, we need to consider their energy, carbon and waste impacts. Therefore this book introduces the concept of embodied carbon, and why it matters. Remembering that as operational energy is improved, embodied carbon, which is the carbon associated with the material extraction, manufacting, transportation, remanufacturing for maintenance, and end of life, becomes prominent. And, it is required to act now, as the majority of embodied carbon is stored in the material production and construction of the buildings. The chapter refers to different methodologies to evaluated it, and show applications in two prefab systems belonging to category 2 and 3 of MMC, posing the question: are lighweight also low-carbon? What is the term of reference? It concludes delineating how potentially futher reduction of carbon could be achieved through localised and nationalised actions.

Finally, it looks at prefab as a possible enabler of circularity. As the necessity to keep materials at their highest value for as long as possible in the construction loop rises, then prefab, capitalising on its modularity, and the use of mechanical connections can enable the reuse of buildings components. This is demonstrated through the analysis of a lightweight steel façade system, through experimental work and lifecycle studies. Concluding that even one reuse of a façade system in a lifetime of a building already has a tangible carbon reduction. Moreover, given the long service life expectancy of many components, enabling multiple reuse should be the way forward.

References

1. Odum HT (2007) Environment, power and society for the twenty-first century: the hierarchy of energy. Columbia University Press
2. IEA International Energy Agency (2024) World energy investment 2024. https://iea.blob.core.windows.net/assets/d829545d-fab6-4c98-b266-28556d86ce8d/WorldEnergyInvestment2024.pdf. Accessed on 10 Aug 2024
3. United Nations (2015) Transforming our world: the 2030 agenda for sustainable development. A/res/70/1. [Online]. https://sustainabledevelopment.un.org/post2015/transformingourworld
4. European Commission (2010) Europe 2020: a strategy for smart, sustainable and inclusive growth. Communication from the Commission. Brussels, 3 Mar 2010. [Online]. Available from: http://eur-lex.europa.eu/LexUriServ/LexUriServ.do?uri=COM:2010:2020:FIN:EN:PDF
5. European Commission (2011) Energy roadmap 2050. Communication from the commission to the European Parliament, the Council, the European economic and social committee and the committee of the regions, COM(2011) 885 final. Brussels, 15 Dec 2011. [Online]. Available from: http://eur-lex.europa.eu/legal-content/EN/TXT/PDF/?uri=CELEX:52011DC0885&from=EN
6. Bourrelle JS, Andresen I, Gustavsen A (2013) Energy payback: an attributional and environmentally focused approach to energy balance in net zero energy buildings. Energy Build 65:84–92
7. Fankhauser S, Smith SM, Allen M, Axelsson K, Hale T, Hepburn C, Kendall JM, Khosla R, Lezaun J, Mitchell-Larson E, Obersteiner M, Rajamani L, Rickaby R, Seddon N, Wetzer T (2021) The meaning of net zero and how to get it right. Nat Clim Chang 12:15–21
8. Ascione F, Bianco N, De Masi RF, De Rossi F, Vanoli GP (2016) Concept, design and energy performance of a net zero energy building in Mediterranean climate. Procedia Eng 169:26–37
9. Lane L, Power A, Provan B (2014) High-rise hope revisited. The social implications of upgrading large estates. Rockwool, CASE report 85, October 2014
10. Webber P, Gouldson A, Kerr N (2015) The impacts of household retrofit and domestic energy efficiency schemes: a large scale, ex post evaluation. Energy Policy 84:35–43

11. Iuorio O, Negro P (2020) A roadmap for sustainable integrated retrofit of concrete buildings. In: Proceeding of the SAFESUST2—SURECON workshop
12. Formisano A, Vaiano G (2021) Combined energy-seismic retrofit of existing historical masonry buildings: the novel "DUO system" coating system applied to a case study. Heritage 4:4629–4646. https://doi.org/10.3390/heritage4040255
13. Preiser FW (1995) Post-occupancy evaluation: how to make buildings work better. Facilities 13(11):19–28
14. Zimring CM, Reizenstein JE (1980) Post-occupancy evaluation. Environ Behav 12(4):429–450. https://doi.org/10.1177/0013916580124002
15. University of Westminster (2004) Guide to post occupancy evaluation
16. Oseland N (2024) A practical guide to post-occupancy evaluation and researching building user experience. Routledge
17. BUS (2021) Building use studies. Available at https://www.designingbuildings.co.uk/wiki/File:Architecture-2_large_.jpg
18. Iuorio O, Wallace A, Simpson K (2019) Prefabs in the North of England: technological, environmental and social innovations. Sustainability 11(14):3884
19. ISO 14040 (2006) Environmental management-life cycle assessment—principles and framework. International Organization for Standardization: Geneva, Switzerland
20. Iuorio O, Gigante A, De Masi RF (2023) Life cycle analysis of innovative technologies: cold formed steel system and cross laminated timber. Energies 16:586. https://doi.org/10.3390/en16020586
21. Iuorio O (2009) Cold-formed steel housing. Pollack Periodica 2(3):97–108
22. One Click LCA Software. Available online: https://www.oneclicklca.com. Accessed on 5 June 2021
23. UK Green Building Council (UKGBC) (2021) Net zero whole life carbon roadmap, Technical Report. https://ukgbc.org/wp-content/uploads/2021/11/UKGBC-Whole-Life-Carbon-Roadmap-Technical-Report.pdf
24. Moschen-Schimek J, Kasper T, Huber-Humer M (2023) Critical review of the recovery rates of construction and demolition waste in the European Union—an analysis of influencing factors in selected EU countries. Waste Manage 167:150–164
25. European Parliament, Council of European Union (2008) Directive 2008/98/EC of the European Parliament and the Council of 19 November 2008 on waste and repealing certain directives. European Union, Brussells, Belgium
26. Fereydooni EA, Khodabakhshian A, Iuorio O, Re Cecconi F, Daniotti B (2024) A new building information modelling-based approach to automate recyclability rate calculations for buildings. Buildings 14
27. Kouvara A, Priavolou C, Ott D, Scherer P, van Zyl-Bulitta VH (2023) Circular, local, open: a recipe for sustainable building construction. Buildings 13:2493
28. Santos P, Cervantes GC, Zaragoza-Benzal A, Byrne A, Karaca F, Ferrández D, Salles A, Braganca L (2024) Circular material usage strategies and principles in buildings: a review. Buildings 14:281
29. Nexxi. https://www.nexii.com/design-for-deconstruction-nexii-circular-economy-case-study/. Accessed on 11 Aug 2024
30. Kitayama S, Iuorio O (2023) Disassembly and reuse of structural members in steel-framed buildings: state-of-the-art review of connection systems and future research trends. J Archit Eng 29(4):03123006
31. Iuorio O, Kitayama S (2024) Experimental study on the feasibility of disassembling and reusing lightweight façade wall systems. In: CESASE 2024 midterm conference of CircularB implementation of circular economy in the built environment. http://hdl.handle.net/11311/1265638
32. Kitayama S, Iuorio O (2023) Using bolted connections for the construction, de-construction and reuse of lightweight exterior infill walls: experimental study. J Arch Struct Constr 4:123–133. https://doi.org/10.1007/s44150-024-00108-0

33. Kitayama S, Iuorio O, Josa I, Borrion A, Black L (2024) Determining the carbon footprint reduction of reusing lightweight exterior infill walls: 1 A case study of a school building in the United Kingdom. J Clean Prod 409:143061
34. Royal Institute of Chartered Surveyors (RICS) (2017) Whole life carbon assessment for the built environment, first ed. RICS, London, UK
35. Gibbons OP, Orr JJ (2022) How to calculate emboided carbon. IStructE Ltd

Chapter 5
Emblematic Prefab Systems

Abstract This chapter presents prefab cases for the construction industry. It does not intend to show them per se, but to articulate how research and development have shaped them, so they can serve as a reference for future development of the construction industry. Therefore, for each case studies, it discusses the historical development of the company, and the pivotal role that research and development have played in the system's evolution. The chapter provides essential information about the systems, including notable examples, demonstrating their real-world use and effectiveness. The selected cases are ordered according to their prefabrication complexity and primary material adopted, offering insights into the manufacturing and assembly challenges and solutions. Examples are categorised also according to their applicability as new constructions or if they have also been developed for retrofitting existing buildings. This structured approach allows for a systematic comparison and understanding of the technological advancements and applications in the field of MMC.

Keywords Construction industry · Building information · Technological system · Embodied energy · Robotic

5.1 WikiHouse

5.1.1 History

WikiHouse started as a start-up in New Zealand, with the aim of developing solutions for post-earthquake housing, after the 2011 Christchurch earthquake. Their initial intent was to build up a system that could be easy deployable as a portacabin, but have the performances and the quality of fully functional buildings that could be passed on for generations and growth with them. Since then the WikiHouse platform has developed in a variety of countries, including UK, Germany, France and USA. The company's overarching objective aligned since its inception with the transformative shifts observed in digital fabrication, particularly within the context of web-enabled

O. Iuorio, *From Mass Prefab to Mass Customization*,
PoliMI SpringerBriefs, https://doi.org/10.1007/978-3-031-73327-7_5

platforms fostering open information exchange. Embracing the ethos of digital fabrication and leveraging advanced manufacturing technologies, WikiHouse aimed to enhance efficiency, customization, and accessibility of housing. WikiHouse ended up providing a platform system to give users worldwide a place to browse and even share prefabricated home ideas. It draws inspiration from conventional construction and flat-pack furniture, offering designs for the main construction components and an assembly instructions manual for each house and structure. Their strategy is a reaction to the reality that not every building type is suitable for every location; while, on the contrary, a design can be adapted for many purposes. Therefore, Wiki prefab construction method enables a large degree of customization to better fit a variety of settings and requirements.

5.1.2 Research and Development

Wikihouse researched developing a system that could be fully digitally designed in a platform that could be shared across different places and countries and be realized within pop-up factories that can fabricate the Wikihouse system through the use of CNC machines. The intent was, since its inception, to democratize construction. In this way, every pop-up factory or client can use the system packages developed by WikiHouse, and make any necessary modifications to fit the specific purpose, but with the requirement to then re-share within the platform the new solution. All that with the intent to grow knowledge, diversify ownership, and distribute fabrication. Crucially, residents are empowered to choose between various options for the construction process. They may opt to engage a professional contractor for the construction of their homes within a stipulated timeframe or alternatively embark on a do-it-yourself (DIY) approach, thereby embodying a spectrum of engagement possibilities ranging from professional assistance to self-directed construction endeavors. This framework not only fosters individual agency and creativity but also promotes a sense of ownership and community engagement within the residential development.

Wikihouse' goal for the future is to further enhance the hardware development system, striving for maximum efficiency, affordability, and ease of assembly. This will be achieved through a comprehensive and accessible online software platform.

5.1.3 The System

WikiHouse has developed two main systems, a first one called WREN, which is based on a wooden portal frame, and it is made of an entirely integrated group of small components that lock together. Two-meter-long sheets of lumber can span up to four meters when offset joints are used in low-stress locations. This method works incredibly well for one-story buildings, but it has limitations when it comes to higher ones.

Utilizing the insights gained from creating WREN, a more user-friendly program has been developed that serves as a toolbox for clients, i.e. as a repository for tools that clients can employ. This second platform system is called Blackbird. The interior architecture of the structures is highly adaptable, with Blackbird offering enhanced functionality also to accommodate wheelchair users. Blackbird allows more flexibility in adding and deleting components, and while everything is interconnected, it is not as complex as the WREN system.

WREN is a portal frame assembly system, every part is different and special items are required for each project (Fig. 5.1). In comparison, Blackbird is post-and-beam style construction, allowing for more freedom in adapting the design without the need to redraw or recalculate. WREN has a limit of 4.5-m spans, Blackbird allows for 5.5-m, anything smaller would result in small-sized rooms and anything larger would be reserved for special cases, such as double garages.

Recently, WikiHouse developed the Skylark block, which is type of cassette system, with a thickness up to 250 mm, for the integration of thermal insulation. The cassette system is completely CNC fabricated starting from flat OSB panels, that are jointed by jigsaw joints [1, 2].

The construction system is well suited for small-scale structures such as garden studios and backyard offices, and small-scale development, such as "pocket sites". Characterized by a lightweight kit of components, this system offers facile installation via narrow access routes, rendering it well-suited for such applications. Its cost predictability and ease of self-installation further enhance its appeal, aligning favorably with the requirements of individuals seeking to develop outbuildings or extensions.

Unlike conventional large-scale developments or volumetric housing approaches, this system mitigates access constraints while affording sufficient flexibility and customization to navigate planning constraints and align with local architectural styles and vernacular architectures. Consequently, it emerges as a versatile solution capable of catalyzing sustainable development on underutilized parcels of land, thereby addressing pressing housing needs and fostering community resilience.

The foundations represent the most costly and unpredictable aspect of construction. Ground engineers determine the required foundation type based on varying ground conditions at each site. During construction, unforeseen factors like air pockets, pipelines, or fossils in the soil can significantly disrupt project timelines and inflate costs, unlike other project components that can be reliably forecasted and managed.

5.1.4 Emblematic Examples

The WikiHouse initiative in the Netherlands is spearheaded by a consortium known as Woningbouwatelier. Their endeavors culminated in the development of a residential site comprising 27 houses situated in Almere, located south of Amsterdam. Notably, this project epitomizes a collaborative effort between the municipality of Almere

Fig. 5.1 Example of cassette systems developed by Wikihouse: for a pavilion in Milton Key where typical Skylark cassette blocks are used (right); example of a cassette framing system developed for the deployment of shelters in Ucraina (left)

and the local community, with the former assuming a leadership role in its facilitation. Fundamentally, the project operates on the premise of serviced plots, whereby residents purchasing these plots are afforded a degree of autonomy in determining the architectural characteristics of their respective dwellings within the prescribed parameters of the WikiHouse system.

In essence, the company's aim encapsulates a multifaceted approach characterized by the convergence of digital fabrication methodologies and open information exchange paradigms. By harnessing the power of digital technologies and cultivating a culture of transparency and collaboration, the company endeavors to catalyze a paradigm shift within its industry. This entails not only optimizing manufacturing processes but also democratizing access to knowledge and resources, ultimately fostering a more inclusive, agile, and sustainable ecosystem of production and innovation.

5.2 LeapHomes

5.2.1 History

Leaphomes is a start-up, born starting from architectural and engineering know-how rather than from a production line. Hence, it has grown independently from a supply chain of materials or technologies.

They conduct their first design experiments to develop construction systems for very particular contexts, like extreme climates and high altitudes. Therefore, they initially designed and fabricated the system for all the places where it is usually not possible to work traditionally, not even with traditional prefabrication where the logistics can be a limitation.

The first projects were all carried out with the special transport of components to the site, like by airplanes, as in the case of the eco-hotel in Russia on the Caucasus at 4000 m above sea level, a context that forced a different approach, much more similar to that of an industrial product. These special requirements determined the need to draw on extremely diversified materials and technologies for a problem of weight and volume reduction, to have macro-elements placed on site according to a plug-and-play logic of perfectly functioning macro components.

That was an opportunity for great research and experimentation, which was coupled with very precious hyper-specialist skills. The requirements led to experimentation with the most advanced composites of various nature, with aerospace insulators, and assembly systems. Subsequently, the acquired know-how was transferred to civil prefabrication.

5.2.2 Research and Development

Leaphouse began to develop an alternative construction system that drew on all the most advanced industrial capabilities, from metalworking to digitalization of processes both at the project and product level as well as in the production of the components. Drawing on extensive research and specific technical requirements, the company initiated the development of an advanced system of prefabricated components, designed to enable entirely mechanical and dry on-site assembly. This approach ensures comprehensive control over all aspects of the building's functionality, encompassing structural elements, the building envelope, interior fit-outs, and integrated services, without compromising design integrity or differentiation between these components.

5.2.3 The System

The developed system has a large series of constant components, having the double integrated function of acting as structural components and casing elements, and designed and fabricated to ensure the maximum possible performance, even going beyond current regulations. The system is optimized thanks to the application of an integrated approach that allows for eliminating the redundancies and using materials that tend to have the highest possible physical and mechanical performances, following a strongly industrial approach.

The product is the component, specifically, the systemic component which allows the same compositional design freedom that any other structural system.

The Leaphome system is therefore conceived as a Meccano, with highly engineered components, developed to ensure the highest performance, and which allow for broad design freedom. The structural and envelope materials vary depending on the application. However, often, the system integrates a patented opaque prefabricated element that couples a semi-structural composite wood core, with a variety of insulations, including also cork, multiple air chambers, and heat-reflective insulators. The element is then completed with a large range of finishes to respond to customization. With an average of a 25 cm opaque component, the system can achieve passive energy efficiency. The elements are then assembled primarily through wood-wood joints, obtained by coupling the wooden elements, which facilitates implementation in terms of precision and positioning.

The final system, therefore, ignores the specialization of the workforce, thus solving the problem of the (de)skilled workforce. The Meccano system is designed to simplify maintenance throughout the building's lifespan. This is achieved by leveraging both digital and physical documentation of each component and service. Since every element is already recorded and classified, replacements can be made easily, either with identical or upgraded components, without needing additional information at a later stage. This ensures that maintenance and upgrades can be performed seamlessly, with all necessary data readily available.

Therefore Leaphome's strength lies in the study of the design for manufacturing and assembly (DfMA) process.

5.2.4 Emblematic Examples

The first building developed by Leapfactory is certainly of great interest, for the extreme conditions in which has been deployed. It is an off-grid settlement at 4000 m above sea level, Mount Elbrus, Kabardino Balkaria. It was installed in 3 days with 38 helicopter rotations, starting from the "base camp" at 1800 m above sea level and 10 km horizontal distance. It comprises 5 buildings, equipped with heating, lighting, VMC, bathrooms and wastewater purification, and also included snow melting to obtain drinking water.

One of the systems developed by LeapHome is called the Frame. The first Frame house, built in 2017, did not have any piece weighing more than 50 kg, which means that theoretically according to the law, it could have been moved completely by hand. The peculiarity of this house was also to be deconstructable, indeed, the house was after some months dismantled, resold, and re-assembled at 1800 m above sea level at Chamois, in the Val d'Aosta region. The system also allowed some interior modifications to respond to the new client requests. Casaclima Bolzano, a consortium company specializing in sustainable construction supply chain, energy efficiency, and intelligent technologies, tested the building performances of a frame house built in Bergamo, and by a comparative analysis with other traditional systems, demonstrated a higher level of performance of the Leaphomes Frame house.

The system itself was born with small and simple typologies and has progressed towards applications to retrofit existing buildings and later to large urban buildings, as in the case of the under-development large-scale development in Milan, named Mind A good example of a retrofit application done to revitalize an abandoned rural buildings of the Alpine villages (Figs. 5.2 and 5.3).

5.3 Lightweight Steel Systems by Etex

5.3.1 History

Etex has headquarters in Belgium and was born from a family business in 1905. Since, then, it has become world-spread over Europe, North America, Asia, and Africa. Etex distinguishes itself from many other companies in developing both products and building systems, ranging from MMC categories 6, 5, and 2. Initially, Etex pioneered fiberboards and gypsum building components, today, they articulate their production across five lines, which are gypsum, cement fiber, fire and high-performance insulations, insulation, and lightweight systems. In particular, bidimensional lightweight steel (LGS) systems having started to become a prominent investment in the last five years, with the increase of interest towards MMC in Europe. To achieve this, they have acquired over the years a range of smaller industrial companies, having growing trajectories and that could complement the in-house expertise.

Fig. 5.2 LeapRus eco hotel—Mount Elbrus, Caucasus, Russia, 2013

5.3.2 Research and Development

Etex has invested a lot in R&D for the development of: the material chemical composition for fiber, gypsum and insulation products; their mechanical and fire resistance properties; water absorption performances; environmental impacts of the products

Fig. 5.3 LeapHome frame, built in Lissone, Milano, Italy, 2017. Adopts the "frame" system

and the system; and recently for the recyclability and circularity of their products and systems.

Indeed, the sustainability of the products and the production and construction process has become a priority for the company. They have proactively worked to control and reduce their products' environmental impacts, by carefully looking at material extraction, operating the gypsum quarries with a conscious eye on the natural surroundings, as well as balancing material chemical compositions, and advancing recycling and reusing. They have indeed strongly invested in reducing waste by optimizing the material, water and energy use. It is certainly worth mentioning their commitment to using wood fibers for fiber-cement products, coming from recycled sources such as used cement paper bags or cuttings and trimmings recovered from cardboard producers. They continuously research for adopting green cement, investing in in-house R&D and collaborating with external research centers and green suppliers. For their lightweight system, made of light gauge steel (LGS) profiles, collaborating with cement and gypsum boards, they invested in structural optimization, which has brought to the adoption of thinner profiles (currently 0.8 mm for single and double storey buildings) saving about 540 tonnes of aluminum annually [3].

Etex tracks and optimized all the material supply chain, and by anticipating, what will become a requirement in Europe, they not only provide Environmental Product Declarations (EPDs) for their products, which support clients in developing life-cycle analysis, but they are also developing Product Passports, which will be instrumental in enabling a circular future.

Among their R&D investments, Etex is, indeed, also investing in advancing the Circular Economy. Raw materials constitute a crucial component of the ecosystem services upon. Addressing the challenges of raw material consumption and waste generation is paramount. Etex developed a roadmap to Sustainability targeting six priority areas. The circularity roadmap outlines their systematic approach to reductions, including circular content in design, achieving zero waste to landfill, implementing take-back and reuse programs, and fostering innovation.

Spotlight of their commitment is their recent involvement in the UKRI Circular Center for Mineral-Based Construction Materials, for which they provided industrial advisory services and case studies for the development of design for disassembly and reuse solutions, and supported scenarios analysis to start quantifying what could be the benefit in terms of embodied carbon (calculated in terms of CO_{2e}) of moving from a linear to circular scenario for the use of Etex façade systems [4]. This study specifically demonstrates that reusing the Etex infill walls beyond the life span of a single building, and be applied for other constructions, could further reduce the embodied carbon by up to 7% for each construction and 11% in cumulative constructions.

To further reduce the footprint of their products, Etex is also planning to adopt XCarb steel, which is made from high levels of recycled steel using 100% renewable electricity in an Electric Arc Furnace [5].

5.3.3 The System

The lightweight steel system division designs and produces lightweight infill wall systems, that in some configuration, is fully integrated with insulation materials.

The Remagin system offers pre-assembled LGS walls and floor systems, which are manufactured in the factory, transported to the site, and then craned into place within either a primary structural frame, or themselves can work as full loadbearing solutions. When the pre-assembled walls have small length (i.e. up to 1.8 m length), they can be manually installed without the need for a crane, for bigger walls Etex has developed an assembly procedure, that can still require no need for scaffolding. The utilization of pre-assembled, panelized walls offers several advantages: (a) enhanced safety for installation workers, as the installation can be performed from within the structure, thereby minimizing the necessity for external work at height; (b) cost reductions due to decreased reliance on mast climbers and scaffolding; and (c) reduced site waste. The pre-assembled LGS walls are usually provided with an external waterproof board when they are used as exterior walls, while they are without boards when they are used as interior walls (and also in this case they can be ordered for load-bearing application or not).

The system is developed for buildings ranging from single storey to fourteen floors. As the required building eight increases, so, the main load-bearing LGS profile changes from C-section with a 0.8 mm thickness to 1 mm thickness and then to omega profiles, that better perform against buckling. The system is well suited for student accommodations, hotels, social housing or condominiums, where the structure's design is conceived for a LGS load-bearing system, and where the all construction process can be optimized by adopting a light prefab system. In countries like United Kingdom, the LGS system can also be found integrated with insulation layers and internal finishing, and externally with the weatherproof boards. To, then be only completed externally according to client requirements.

Etex has also developed a flooring system, made by LGS cassettes, which are primarily filled with low carbon concrete, to provide a good thermal mass, reduce noise issue, which can be typical of light flooring systems, and have a good fire rating.

5.3.4 Emblematic Examples

To understand the scale of MMC, and the speed at which is developing, it has been chosen as emblematic example for Etex, the Carrow Quarter, in Norwich. The building comprises 101 affordable residential units distributed across eight floors. Emphasizing the benefits of LGS and collaborative efforts, the teams from Sigmat, which partner in Etex, and the contractor Carter, innovated a safer construction technique by assembling the balconies at ground level and then elevating the completed components into their designated positions. This approach not only reduced on-site risks but also conferred additional scheduling advantages, ultimately leading to the project's early completion (Fig. 5.4).

Fig. 5.4 Multistorey residential building in Carrow Quarter, in Norwich. Photos courtesy of Etex

5.4 Irondom

5.4.1 History

Irondom is a company established in the South of Italy in 2018, in an area characterized by a double complexity. The area is characterized by a quite old and poorly built environment, in which innovation in construction struggles. The region requires high construction safety standards, as it is an area of high seismic activity, which was strongly devasted by an earthquake (the Irpinia earthquake) in 1980. Therefore, Irondom intended to develop an industrialized construction system with high structural performances in seismic areas while providing indoor comfort and energy efficiency. They developed a construction system based on light-gauge steel frame technology (LSF) that can either assembled on site starting as a stick-built solution or can be preassembled in the factory at the 2D-panel scale.

Their goal is goal is to offer buildings that are adaptable to the continuous evolution of life, sustainable, and aesthetically pleasing.

The company aims to achieve its objectives—enhancing people's lives, and improving the comfort and safety of buildings, making them easy to design, construct, and use—through ongoing innovation and collaborations with leading polytechnic institutions in Italy.

5.4.2 Research and Development

Research and development were instrumental for Irondom as National and European codifications currently do not allow for adopting cold-formed steel (CFS) systems in seismic areas. However, in the region, CFS has already been used for a school's seismic design and construction [6, 7], and was developed according to a design assisted by a testing approach. In agreement with that principle, Irondom developed its industrialized system based on experimental tests at the microscale (connection systems) and mesoscale (wall system). One of their objectives was to adopt a limited number of steel profiles, based on a 100 mm height C section, that could be used for both wall and floor systems, to simplify connection detailing and contain machine capital costs, which can be significant at the start of any industrial venture. They also decided to adopt a sheathing-braced approach, for lateral and vertical loading, employing a structural OSB panel on one side of the wall (see Chap. 3 and Iuorio [9], for a detailed explanation of the methodological approach, and refer to the company website for more explanation of their experimental studies).

The system was also further developed in the so-called "Duo System" for the energy and structural retrofit of existing buildings. To this end, large numerical analyses were employed to evaluate structural scenario analysis, as well as to study the most appropriate technological stratification to achieve a range of thermal and acoustic performances. So that they currently offer solutions for ventilated and non-ventilated retrofit wall solutions.

5.4.3 The System

Irondom has developed a LSF-based dry construction system, which uses as basis 100 mm CFS profiles, that are braced by OSB panels (either 9 and/or 15 mm) and adopts rockwool for thermal insulation and mineral wool for acoustic insulation. To this basic insulation, further layers of insulated metal alloy panel with a 22 mm rigid wood-cement panel underneath are added to achieve higher thermal performance.

The basic wall system is conventionally finished, towards the exterior with cement-based panels, that can be plastered and painted, as wished, and it is finished on the inner side with a void for services and double gypsum panel. They also provide a ventilated version of the wall system, which employs a thermal insulating layer on the outside, composed of an expanded polyurethane panel, covered with a vapor barrier, a 43 mm ventilation chamber, and can be finished toward the outdoor with either plastering, ceramic or metal panels.

Irondom also has developed a series of floor solutions which are based on the use of a built-up lattice beam, with the upper joist inclined if it has to achieve a required slope. The beams are then connected on the top to a double 15 mm OSB panel and uses a rigid wood-cement panel and a sandwich thermal insulated corrugated metal alloy panel for thermal and finishing purposes. The roof solution is completed towards the inside with 45 mm mineral wool sound-absorbing layer and plasterboard panel. As for the walls, as pallet of finishing are available for the roof system, including ventilated solutions.

The Duo System, developed for the retrofit of existing buildings, represents a significant enhancement for buildings with masonry and reinforced concrete structures. Its seismic improvement and anti-tipping properties are attributed to a frame composed of aluminum profiles and a stiffening panel, which are securely connected using screws. Thermal enhancement is achieved through a dual-layer insulation approach. The first layer, a 45 mm glass wool panel, is placed between the existing building wall and the exterior coating. The second layer, integrated within the 100 mm thickness of the frame, can consist of materials such as hemp, heat-reflective materials, glass wool, or rock wool, tailored to the specific requirements of the project. The system is affixed to the building's perimeter walls using plates and plugs designed for various structural types, with special insulation using thermostops to eliminate thermal bridges. This installation consistently ensures improved seismic performance, energy efficiency, and aesthetic upgrades.

5.4.4 Emblematic Examples

Among the projects developed by Irondom, it is certainly of interest to look at a rural building, and the recent retrofit of a building in Milan, (Italy). An appropriate example of the Irondom system for new construction is provided by the Rural House built in the Park Taburno Camposalo, completely immersed in an olive grove. The house is designed for provide low environmental impact, waste reduction and energy efficiency, and includes also the recovery of rainwater.

A condominium complex comprising multiple buildings has undergone the installation of a system designed to enhance seismic, energy, and acoustic safety performance, as well as to renovate the facade. The use of dry assembly techniques allowed for the partial removal of the pre-existing clinker cladding solely in areas exhibiting deterioration, thereby minimizing demolition activities and reducing waste management costs (Figs. 5.5 and 5.6).

Fig. 5.5 New construction of a rural building in the regional park Taburno Camposalo

Fig. 5.6 Retrofit the multistorey building "Condominio Opera" in Milan, showing the building before, during and after the façade renovation. Photos courtesy of Irondom

5.5 Automated Architecture

5.5.1 History

Automated Architecture is a spin-out university company, whose genesis can be traced back to a critical examination of conventional modular construction methodologies. As mentioned in the previous chapter of this book, traditionally, modular construction has been synonymous with the operation of large, centralized factories characterized by substantial financial investments, extensive labor requirements, and standardized product outputs. However, this prevailing approach necessitates a confluence of factors including substantial up-front capital investments, consistent project pipelines, and readily available labor pools. Yet, it has become increasingly evident that this model is inherently flawed, evidenced by the recurrent failures and insolvencies within the industry.

Recognizing the limitations and shortcomings of the established modular construction paradigm, the academics embarked on a journey of radical reevaluation. Central to this reimagining was the acknowledgment that the prevailing model is ill-equipped to address the pressing housing crisis. Thus, driven by a commitment to innovation and societal impact, the academic institution (UCL) sought to transcend the confines of traditional modular construction approaches.

This endeavor marked the inception of a transformative spin-out initiative, characterized by a departure from the conventional centralized factory model. Instead, the spin-out ventured towards a decentralized, agile paradigm that prioritizes adaptability, scalability, and efficiency. By challenging the status quo and harnessing cutting-edge technologies, the spin-out endeavors to redefine modular construction, offering a sustainable solution to the housing crisis while catalyzing advancements within the construction industry. The result is a MMC system that revolves around advanced and robotic manufacturing.

5.5.2 Research and Development

AUAR building system is meticulously engineered with automation, particularly having robotic automation, at its core. The conception entails a holistic perspective, envisioning a building system capable of optimal adaptability, agility, and versatility to facilitate scalability. This conceptualization manifests in a building system founded upon modular components adjustable according to a standardized unit of measurement, facilitating seamless scalability for both larger and smaller configurations, all manufacturable using a uniform process. Central to this paradigm is the strategic integration of robotics, maximizing the value addition within the manufacturing process.

This comprehensive building system not only addresses structural elements but also establishes a standardized framework guiding various aspects including cladding, insulation, and interfacing with other standardized building products. Consequently, it fosters diverse housing typologies, accommodating varying needs while maintaining consistency in the manufacturing process.

This paradigm shift alleviates bottlenecks in construction by negating the necessity to approach each project or manufacturing process as a unique endeavor. Rather, it enables the establishment of micro-factory setups wherein housing production can be efficiently executed with the utilization of one to several robots. Moreover, the modular nature of these factories allows for scalability, facilitating adjustments in accordance with project pipelines, and thereby mitigating planning risks. Positioned as a low capital expenditure with minimal operational costs, AUAR adaptable setup caters specifically to the needs of the construction pipeline. Such innovations are marketed to contractors and other stakeholders within the value chain interested in embracing automation technologies.

To achieve this, the work embeds years of research and development in the field of robotic planning, manipulation, as well as architectural production. A theoretical framework of AUAR research can be found in "Robotic Building: Architecture in the Age of Automation" [8]. Their research looked at technological innovation, operational efficiency, and sustainability principles. By harnessing cutting-edge automation technologies within a micro-factory framework, AUAR ended up offering a system solution that addresses many inefficiencies plaguing traditional construction methodologies, while allowing for clients to be involved in the project development thanks also to the design tools they developed, which allow through a design configurator people to easily design their own houses.

5.5.3 The System

The building system is based on modular wooden construction, whose capabilities have been defined by considering its scalability across various architectural contexts. The building block once its completed can be the internal face and can be treated in a variety of ways to make it specific to people's preferences. While they use standardized approach for interfacing with cladding and vapour barrier membranes, leaving open to a range of possibilities for the cladding, which could be ceramic, timber cladding, composites, or many more off-shelf products. The basic building block itself is designed and conceived to achieve passive house standards independently from the adopted cladding.

The developed system accommodates a spectrum of building scales, ranging from backyard micro-dwellings to six-stories multifamily structures. This deliberate limitation aligns with the inherent complexities associated with timber construction beyond the six-floors threshold, and the complex regulatory constraints that would incur beyond that limit. Thus, the system is optimized to operate within this range, and it suitable to be used to densify urban infill sites.

Moreover, the system demonstrates a remarkable capacity for contextual adaptation, exemplified by its deployment in disparate countries such as the UK and Belgium. The adaptability of the system extends beyond mere spatial considerations to encompass nuanced variations in dimensions, layout configurations, and typologies without necessitating fundamental alterations to the underlying building system. This flexibility is akin to a "pixel building" paradigm, where the system serves as a versatile framework capable of realizing a diverse array of architectural forms.

While certain localized adjustments may be requisite to address climatic considerations, such as the incorporation of vapor barriers, these modifications do not undermine the fundamental integrity of the building system nor disrupt the manufacturing process. Thus, the system emerges as a potent enabler of architectural diversity and spatial adaptability, offering a robust solution that transcends geographical boundaries while remaining attuned to local contexts and regulatory requirements.

The innovation underlying AUAR approach to construction automation represents a significant departure from prevailing methodologies within the industry. While certain companies have made strides in robotic assembly, their focus typically revolves around automating the construction of individual wall panels. However, it's imperative to underscore that this approach necessitates substantial infrastructure investment and remains inherently resource-intensive.

What sets AUAR innovation apart is the holistic micro-factory model, which represents a paradigm shift in construction automation. Unlike existing practices, their approach transcends the limitations of fragmentary automation by encompassing the entire construction process within a compact, agile manufacturing facility This approach not only streamlines production but also minimizes resource consumption and enhances operational efficiency.

5.6 Reframe Systems

5.6.1 The History

The company aims to deliver net-zero carbon housing at scale by utilizing design automation, modular building components, and digital manufacturing within robotic microfactories. The CEO and the main founders come from Amazon Robotics, and they bring their manufacturing approach to the construction sector to respond the question of "how to design, and manufacture net-zero housing which can be affordable for all". The funders leveraged their expertise in scaling automation solutions to address carbon emissions and create societal impact. Collaborating with Eclipse Ventures, they developed a thesis on the construction of net-zero-carbon homes using modular construction within flexible microfactories, aiming to enhance speed and cost-efficiency. Following the validation of the business case, they established Reframe Systems with initial funding from Eclipse Ventures and Fundamental Ventures.

In the attempt to tackle climate emergency and housing affordability, while learning from some failures of large-scale housing production, Reframes has started developing a building platform and production system that enable mass customization. The platform is rooted in the principles of flexible automation, using readily deployable robotic work cells, that are low-cost, reconfigurable, and built to scale up or down in distributed and decentralized networks. Reframe Systems uses automation and augmentation to build prefab wall panels and full houses, for which they transferred the knowledge gained in Amazon to assemble prefab housing system in a space and time-efficient way.

5.6.2 Research and Development

Reframes's funders experience with robotic systems at Amazon provided valuable insights, as they utilizes widely adopted robotic arms commonly seen in the automotive industry. This familiarity with robotic technology has been instrumental in the successful development and deployment of their innovative prefabrication system.

Their first factory is only about 1500 m^2 and they are still able to build roughly 9000 m^2 of product a year out of it. They have changed the way to construct prefab systems, by building their walls in a vertical position. Which means, that whilst most of the other prefab builders, producing panelised systems, build their walls horizontally flat and then turn them vertically, Reframe makes the walls already vertical, so that in a limited industrial space they can build up full volumes.

They use, at the moment, a single robotic arm to frame full walls. They are using vision algorithms to actually identify wall studs, pick, play an plug them in the required configuration, instead of programming the robot. This means they use a dynamic environment, in which the robot adjusts its motion planning through continuous learning.

The funders have taught the robot to execute its motions. This allows the robot to keep re-executed different paths and different plans. The benefit of this approach is that in case of any change in the wall configuration, like a change in the header type, fixing procedure, geometry, dimensions, and so on, there is no need to re-program the entire robotic system to do it. The Reframes Systems platform is developed so that the instructions are directly transferred from the CAD system into the robot. Therefore, in case of any change in the CAD document, the robot can directly adjust and react to those changes in geometries. That is a very unique and novel way to do robotics for manufacturing, which learns directly from other automation industries.

Reframe Systems plan is to soon have two robots framing and sheathing the walls, and then they will build up a mobile platform for allowing a more agile automated environment, and take care of drywall, finishes, and paint applications. In their roadmap, Reframe is planning to have about four robotic work cells for framing and sheathing, and up to six mobile robots for drywall, finish taping, and finishing applications, and that would allow them to build up to about 100 homes per year. While people with iPads will instruct and check the advancements.

Reframes is investing into three continuous areas of R&D, which look at: (1) the development of a new building platform, which would no longer depend on local materials; (2) computational design, to develop algorithms able to translate the entire exercise of concept design, digital design, detailed design and transfer to manufacturing instructions, (i.e. go from pixel to parts completely through digital computation); (3) automation to simplify the robot programming process, and make robots more efficient, while also improving the vision feedback system.

5.6.3 The System

Reframe primarily targets single and double-family houses, looking at the triple deckers as the primary housing typology applicable in North America. They primarily use a wooden system, similar to the classic American platform frame, which is realized by using local material, bought from local lumberyards and Home Depots. This supply chain allows Reframes to control the stage's material costs.

Reframes uniqueness is in framing walls vertically rather than horizontally, and assembling the volumetric units in their vertical position. This approach eliminated the need for spatial manipulation required in horizontal framing and subsequent vertical flipping, thereby streamlining the process. This strategic decision brought about additional advantages.

To do that, they have developed a robotic work cell capable of framing entire walls (Fig. 5.7). The cost of constructing this robotic system was approximately $120,000, and it has the capacity to produce between 9 to 15 linear meters of wall per hour. This makes Reframes' system highly cost-effective.

Fig. 5.7 Robotic platform for vertical assembly of wooden prefab walls by reframe systems. Photo courtesy by Vikas Enti

Reframe is positioned to significantly impact current workforce challenges in the industry by leveraging automation and robotics. This strategic approach addresses design-related issues, primarily by enhancing team productivity and mitigating the labor shortage through automation.

Furthermore, the adoption of automation can tackle retention challenges faced by architects and engineers, who often seek more career opportunities and greater engagement. By integrating design and automation, the Reframe Systems create new avenues for career advancement, enabling professionals to develop their skills and pursue more fulfilling roles.

This transformative journey, blending creativity with technological innovation, invigorates workplace engagement. Embracing this shift has the potential to rejuvenate employee motivation and foster a dynamic, evolving workforce, inspired by the continuous advancements in construction industry.

References

1. Goudie G (2023) Experimental testing of plywood & OSB 'jigsaw' joints for structural characteristics. Dissertation. Accessed at https://www.wikihouse.cc/research/experimental-testing-of-plywood-osb-jigsaw-joints-for-structural-characteristics
2. Granello G, Yang E, Zhang Z, Ogunkuade O (2023) Experimental testing of OSB Skylark floor beams. Testing report. Malaga. https://www.wikihouse.cc/research/experimental-testing-of-osb-skylark-floor-beams. Accessed on 1 June 2024
3. Etex (2023) Annual report. Sustainable Solutions for you. Available at https://www.etexgroup.com/en/investors/annual-reports/. Accessed on 21 May 2024
4. Kitayama S, Iuorio O, Josa I, Borrion A, Black L (2024) Determining the carbon footprint reduction of reusing lightweight exterior infill walls: 1 A case study of a school building in the United Kingdom, under review. J Clean Prod
5. ArcelorMittal Europe—Flat steel. Towards carbon neutral steel, 2023. Available https://corporate.arcelormittal.com/media/elwjci24/xcarb_flatproducts_brochure.pdf. Accessed on 23 May 2024
6. Iuorio O, Fiorino L, Landolfo R (2014) Testing CFS structures: the new school BFS in Naples. Thin-Walled Struct 84:275–288

7. Fiorino L, Iuorio O, Landolfo R (2014) Designing CFS structures: the new school BFS in Naples. Thin-Wall Struct 78:37–47
8. Retsin G, Jimenez M, Claypool M, Soler V (2019) Robotic building. Architecture in the age of automation, DETAIL.https://doi.org/10.11129/9783955534257
9. Iuorio O (2009) Design Procedures for Cold-Formed Steel Housing in Seismic Area. PhD Thesis, University of Gabriele D'Annunzio

SPRINGER NATURE

GPSR Compliance

The European Union's (EU) General Product Safety Regulation (GPSR) is a set of rules that requires consumer products to be safe and our obligations to ensure this.

If you have any concerns about our products, you can contact us on ProductSafety@springernature.com

In case Publisher is established outside the EU, the EU authorized representative is:

Springer Nature Customer Service Center GmbH
Europaplatz 3
69115 Heidelberg, Germany

The manufacturer's authorised representative in the EU is Springer Nature Customer Service Centre GmbH, Europaplatz 3, 69115 Heidelberg, Germany. If you have any concerns regarding our products, please contact ProductSafety@springernature.com

Printed and bound by CPI Group (UK) Ltd, Croydon, CR0 4YY
30/09/2026
02225474-0002